Offshore Marine Operations

Offshore Marine Operations

Edited by Adam Hassan

Cover image: by author

First published 2022

By Magellan Maritime Press Ltd

Millbank Business Centre, 1 William Street, Southampton, Hampshire, SO14 5QH, United Kingdom

© Magellan Maritime Press Ltd

British Library Cataloguing in Publication Data

A catalogue record for this book is available from the British Library

ISBN 9781739774394

Magellan Maritime Press Ltd
Millbank Business Centre, 1 William Street, Southampton, SO14 5QH, Hampshire, United Kingdom
+44 (0)23 81947782
contact@magellanmaritimepress.com
www.magellanmaritimepress.com

Typeset in Sabon

Preface

This book is an adaptation of the *Guidelines for Offshore Marine Operations*, which were published as Revision 0611-1401 on 06 November 2013. References to the *Guidelines* or *Guidance* is made to direct the reader to the appropriate section of the original publication. The objective of this publication, therefore, is to provide said guidance regarding the best practices which should be adopted to ensure the safety of personnel on board all vessels servicing and supporting offshore facilities, and to reduce the risks associated with such operations. It particularly relates to the following activities:

1. Operations of offshore facilities.
2. Operations of vessels.

Whilst the best practices summarised in this document primarily reflect those adopted in the North-West European Area, the author recognises that the guidance may just as easily apply outside this region and that many, if not all, of the recommendations included do indeed have global relevance. Where it has been possible to make recommendations relating to operations outside its core area without diluting the original objectives these have been included. It is recognised, however, that in certain circumstances local or company-specific requirements may exist. In this event the guidance contained herein should be read in the context of such requirements and interpreted accordingly. To facilitate common practices on a global basis, where necessary, the guidance, together with included reporting forms, should be used as the basis for preparing procedures for local practices.

Adam Hassan
Editor

Editor's Note

Any references in this document to facility manager, OIM, master, base manager or any other person in authority should be interpreted as also including their nominated deputies. In the context of this document "operator" refers to the party responsible for the management of petroleum activities on behalf of the licensees. In the context of this guidance "owner" relates to the party responsible for the management of one or more OSVs and includes those operating tonnage managed on behalf of others. References to further information from a wide range of sources, both public and private, have been included in this document.

In identifying any references to be included the following principles have been adopted:

1. The information included is generally accepted to represent best industrial practice.
2. The information included may be used from time to time as basis of design or in marine operations manuals.
3. The information included is subject to regular and rigorous peer review, being updated as required.
4. Any references in this document to the masculine gender relate equally to the feminine gender and should be interpreted accordingly.
5. The information is included in the public domain, preferably in an electronic format and free of charge.
6. There is no commercial benefit to the source of the information as a result of its inclusion in this publication.

Any references in this publication to the masculine gender relate equally to the feminine gender and should be interpreted accordingly.

Acknowledgements

I would like to thank the many individuals who have provided me with assistance and material during the writing of this book. To the many colleagues and friends who have answered numerous queries and added their wealth of experience, I extend my grateful thanks and gratitude.

Adam Hassan

Contents

Figures

Tables

Abbreviations

24 / 7	24 hours per day, 7 days per week
AH	Anchor handling
ABS	American Bureau of Shipping
AHTS	Anchor Handling Tug Supply Vessel
AHV	Anchor Handling Vessel
BP	Bollard Pull
CBP	Continuous Bollard Pull
CMID	Common Marine Inspection Document (sponsored by IMCA)
COLREGS	International Regulations for Prevention of Collisions at Sea, 1972
CoS	Chamber of Shipping (trade association representing owners and operators of UK-based shipping companies)
COSHH	Control of Substances Hazardous to Health
COWSP	Code of Safe Working Practices for Merchant Seamen
DC	Daughter Craft
DGPS	Differential Global Positioning System
DMA	Danish Maritime Authority
DNMI	Det Norske Meteorologiske Institut
DNV	DNV Det Norske Veritas
DP	Dynamic Positioning
DPO	Dynamic Positioning Operator(As defined by IMCA, MTS, etc.)

DSA	Danish Shipowners' Association
DSV	Diving Support Vessel
ERRV	Emergency Response and Rescue Vessel
ERRVA	Emergency Response and Rescue Vessel Owners' Association
ETA	Estimated / Expected Time of Arrival
ETD	Estimated Time of Departure
FMEA	Failure Mode and Effect Analysis
FPSO	Floating Production, Storage and Offloading Unit
FRC	Fast Rescue Craft
GLND	GL Noble Denton
GOMO	Guidelines for Offshore Marine Operations
GPS	Global Positioning System
HAZID	Hazard Identification
HAZOP	Hazardous Operations (Assessment)
HF	High Frequency (Radio)
HIRA	Hazard Identification and Risk Assessment
HSSE	Health, Safety, Security and Environment
Hs, Hs	Significant Wave Height
HSE	Health and Safety Executive (UK Government Agency)
HSSE	Health, Safety Environmental and Quality (Management) (generic term used throughout this publication)
JAG / TI	Joint Action Group / Temperature Indices
IACS	International Association of Classification Societies
IADC	International Association of Drilling Contractors
IBC	International Code for the Construction and Equipment of Ships carrying Dangerous Chemicals (IBC Code)
ICS	International Chamber of Shipping
ILO	International Labour Organisation
IMCA	International Marine Contractors Association (trade association for marine contractors engaged in supporting the offshore industry or similar bodies)
IMDG	International Maritime Dangerous Goods Code
IMO	International Maritime Organisation
IMPA	International Marine Pilots' Association
INLS	International Noxious Liquid Substances Code
ISM	International Safety Management Code
ISPS	International Ship and Port Facility Security Code
JSA	Job Safety Analysis
KATE	Knowledge, Ability, Training and Experience
LRS	Lloyds Register of Shipping

MARPOL	International Convention for the Prevention of Pollution from Ships (MARPOL) (IMO Convention 1973 and as subsequently amended)
MBL	Minimum Breaking Load
MCA	Maritime and Coastguard Agency
MF	Medium Frequency (Radio)
MGN	Marine Guidance Note (issued by the MCA)
MLC	Maritime Labour Convention (ILO Convention 2006)
MOC	Management of Change (Process)
MODU	Mobile Offshore Drilling Unit
MOU	Mobile Offshore Unit
MSC	Maritime Safety Committee (IMO Committee)
MSDS	Material Safety Data Sheet
MSN	Merchant Shipping Notice (issued by the MCA)
MTS	Marine Technology Society
MWS	Marine Warranty Surveyor
NMA	Norwegian Maritime Authority (replaces NMD)
NMD	Norwegian Maritime Directorate
NOGEPA	Netherlands Oil and Gas Exploration and Production Association
NOROGA	Norwegian Oil and Gas Association (replaces OLF)
NSA	Norwegian Shipowners' Association
NWEA	North West European Area
OG UK	Oil and Gas UK (trade association for UK Offshore operators and support contractors)
OCIMF	Oil Companies' Industry Marine Forum (trade association for major oil companies engaged in marine activities)
OIM	Offshore Installation Manager
OLF	Oljeindustriens Landsforening (Norwegian oil industry association, replaced by NOROGA)
OMHEC	Offshore Mechanical Handling Equipment Committee
OOW	Officer of the Watch
OSV	OSV
OVID	Offshore Vessel Inspection Database (sponsored by OCIMF)
PCP	Permanent Chaser Pendant / Pennant
PIC	PIC (of a MOU)
PLB	Personal Locator Beacon
PM	Planned Maintenance (System)
PMS	Power Management System
PPE	Personal Protective Equipment
PSA	Petroleum Safety Authority
PSV	Platform Supply Vessel

PTW	Permit to Work
RA	Risk Assessment
ROV	Remotely Operated Vehicle
SBV	Stand-By Vessel
SCV	Small Commercial Vessel Code
SDPO	Senior Dynamic Positioning Operator (as defined by IMCA, MTS, etc.)
SIMOPS	Simultaneous Operations
SJA	Safe Job Analysis
SMC	Safe Manning Certificate
SMPEP	Shipboard Marine Pollution Emergency Plan
SOLAS	International Convention for the Safety of Life at Sea (SOLAS) (IMO Convention 1974, as subsequently amended)
SSV	Safety Stand-By Vessel or Stand-By Safety Vessel
STCW	International Convention for Standards of Training, Certification and Watchkeeping for Seafarers (IMO Convention 1978, as subsequently amended)
SWL	Safe Working Load
TBT	Toolbox Talk
TMS	Tug Management System
UHF	Ultra High Frequency
UKCS	United Kingdom Continental Shelf
UKOOA	United Kingdom Offshore Operators' Association (now Oil & Gas UK)
VHF	Very High Frequency

Definitions

Accident	Undesired event resulting harm to persons, environmental pollution or damage to physical assets.
Adverse weather	Environmental conditions requiring precautionary measures to safeguard the facility or maintain safe working.
Asset(s)	Any infrastructure or equipment associated with offshore production.
Banksman	Person on installation or vessel guiding the crane operator May also be referred to as *Flagman* or *Dogman*.
Base	Quay facilities with logistics support dedicated to petroleum activities.
Base Company or Operator	Owner or operator of a base.
Base Manager	Person responsible for operations on the base.
Blow Off	See *Drift Off*.
Blow On	See *Drift On*.
Bollard Pull	The towing vessel's pull normally specified as maximum continuous pull.
Bridle Towing Arrangement	Two wires or chains of equal length arranged as a triangle that connects the towed object to the vessel towing it.
Catenary Curves	Specification of towline and anchor line curvature for various loads.
Chafe Chain	Short length of chain in way of fair leads to minimise wear on wire or rope bridle components.
Chain Tail	A short length of chain consisting of two or more links.
Charterer	Party hiring marine vessel either on behalf of itself or other interests.

Cherry-picking	Selective discharge of cargo from within the stow.
Competence	Acquisition of knowledge, skills and abilities at a level of expertise sufficient to be able to perform a task to a required standard.
Confined Space	A free entry, non-dangerous space where the relevant risk assessment has identified that under exceptional circumstances there would remain a (remote) possibility for the atmosphere to be adversely affected. Entry and egress routes to such spaces likely to be restricted and controlled by permit.
Coxswain	Generic term for PIC of a small craft.
Dangerous Space	Enclosed or confined space in which it is foreseeable that the atmosphere may at some stage contain toxic or flammable gases or vapours, or to be deficient in oxygen, to the extent that it may endanger the life or health of any person(s) entering that space.
Daughter Craft	Larger fast rescue craft of semi-rigid construction and typically up to 11 metres in length, provided with fixed protection from elements for crew and recovered survivors, capable of being deployed from host vessel for periods of up to six hours.
Dogman	See *Banksman*.
Down Weather	A position on the lee side of an offshore facility or vessel.
Dynamic Positioning	Dynamically positioned vessel (DP Vessel) means a unit or a vessel which automatically maintains its position (fixed location or predetermined track) exclusively by means of thruster force.
Drift Off	Circumstances whereby, in the event of loss of power, environmental forces would result in a vessel moving away from an offshore facility or other navigational hazard.
Drift On	Circumstances whereby, in the event of loss of power, environmental forces would result in a vessel moving towards an offshore facility or other navigational hazard.
Duty Holder	In relation to a fixed installation, this is the operator. In relation to a mobile installation it is the owner.
Emergency Situation	Any unplanned event which may result in harm to persons, environmental pollution or damage to physical assets.
Facility, Offshore	In the context of this document any physical structure on or above the surface of the sea in the vicinity of which marine operations are undertaken. This term includes bottom supported and floating installations, drilling units of all types and other vessels engaged in offshore support operations.
Flag State	Jurisdiction where a vessel is registered.
Flagman	See *Banksman*.
Gog (or Gob) Wire	Wire used to control movement of main tow line when vessel is engaged in towing operations.
Gypsy	Wheel with machined pockets for hoisting chains fitted on a winch.
Hold Point	Stage in any operation at which progress will be assessed to ensure that anticipated objectives at that point have been achieved and that all conditions are favourable for safe continuation of activities. Proceeding past each hold point may require formal acknowledgement in procedures or operational logs.

Hot Work	Welding, burning or flame producing operations.
Incident	Undesired event resulting in damage to assets, equipment or the environment.
Installation, Offshore	Installation, plant and equipment for petroleum activities, excluding supply and standby vessels or ships for bulk petroleum transport. Includes pipelines and cables unless otherwise provided. A structure for exploration or exploitation of mineral resources or related purposes that is, will be, or has been used whilst standing or stationed in water, or on the foreshore or land intermittently submerged.
Inter-field Operations	Operations carried out by vessels between offshore facilities.
J-Chaser	Hook used by anchor handling vessels to *fish* the installation's anchor lines.
Kenter Link	Device for linking two chain lengths.
Lee Side	That side of an offshore facility (or vessel) away from which wind is currently blowing.
Logistics Company	Organisation which, on behalf of its clients, arranges for the transportation of cargo to or from offshore facilities.
Logistics Service Provider	See *Logistics Company*.
Master	Nominated person having command or charge of a vessel. Does not include any pilot.
Mechanical Means of Rescue (Recovery)	Arrangements installed on a Stand-By Vessel to facilitate rescue of survivors from the sea in circumstances where rescue craft cannot safely be deployed or recovered. Proprietary designs include the Dacon Scoop and Sealift Basket.
Mechanical Recovery Device	As for *Mechanical Means of Recovery*.
Mechanical Stopper	Device for temporarily securing chains or wires to facilitate safe connection or release. Proprietary designs include the Karm Forks and Triplex Stopper.
Near-Miss	Undesired circumstance with the potential to cause harm, injury, ill health, damage to equipment or the environment.
Nominated Manager	Nominated persons *in charge* of a specified area or task to be performed.
Non-conformity / Non-compliance	A circumstance where guidelines, regulation or legislation have not been followed.
North-West European Area	Area which includes the north-west European continental shelf and extending 200 miles from any coastline.
Offshore Installation Manager	The person in charge of an Offshore Installation, also known as the facility manager.
OSV	Any vessel involved in supporting offshore activities which is not a mobile offshore unit.
Operating Company / Operator	Party that carries out the management of petroleum activities on behalf of licensees.

Owner	In the context of this document refers to the owner of an OSV. This term may also refer to vessel managers responsible for operating tonnage on behalf of others.
Pear Link	Device for linking two different chain dimensions.
Pendant	Wire hanging permanently attached to the installation used for chasing out anchors.
Pennant Wire	Buoy wire; wire from the seabed up to a buoy on the surface.
Permanent Chaser	Collar through which an anchor chain runs, to which recovery pendant wire is attached.
Personnel Transfer Basket	Equipment utilised for transferring personnel by crane. May also be referred to as a Personnel Carrier.
Piggyback Anchor	Any additional anchor connected to the primary when the latter anchor has insufficient holding capacity.
Pigtail	Short chain or wire with open end links.
Post State	State having jurisdiction over activities in its ports and territorial waters.
Radio Silence	Restrictions of limitations to radio transmissions whilst with a safety zone, usually relating to handling of explosives on the facility.
Recognised classification society	Classification society recognised by IACS to approve vessel design, construction, outfitting and operations.
Redundancy	The ability or possibility of a component or system to maintain or re-establish its function following a failure.
Risk Assessment	A process of assessing risk in any operation.
Safety Delegate	Nominated representative for crew or part of crew or group of workers with regard to health, safety and environmental matters. May also be referred to as Safety Representative.
Safety Zone	Established within a radius extending to distance determined by the relevant legislations beyond the outline of any installation, excluding submarine pipelines.
Sector State	State having special rights and jurisdiction over the development of marine resources within its exclusive economic zone.
Sharks Jaws	See *Mechanical Stopper* above.
Ship Owner	Those responsible for normal vessel management and operation.
Shipper	A person who, as principal or agent for another, consigns goods for carriage by sea.
Significant Wave Height	Average height of the highest one third of the waves over a period of 20 minutes.
Simultaneous Operations	In the context of this document two or more vessels supporting the same or different operations within the safety zone around an offshore facility.
Socket, Wire Rope	Any manufactured end termination fitted to the end of a wire rope to facilitate the connection of other rigging elements.
Spooling Gear	Arrangement to guide wire onto drum.
Stand-By Vessel	Older term for Emergency Response and Rescue Vessel.

Stand-By Vessel	Any vessel mobilised to provide response and rescue support at one or more offshore facilities. Such support will primarily involve the rescue of personnel from the sea and their subsequent care. It may also include fire fighting. May also be referred to as *Emergency Response and Rescue Vessel*, *Safety Stand-By Vessel* or *Stand-By Safety Vessel*.
Stern Roller	Large roller on the stern of an anchor handling vessel to facilitate the recovery or deployment of moorings or other equipment.
Stinger	In the context of this document the pennant installed on the crane's hook to facilitate the safe connection and release of the lifting rigging on any item of cargo. A suitable safety hook will be fitted to the lower end of the pennant.
Supply Chain	Base or base company, vessel or ship owner, installation or operating company.
Supply Service	Supply and or receipt of goods to or from offshore facilities.
Surfer	Small or medium sized high speed craft used for transportation of personnel or light cargoes in benign areas of operations. Foredeck design is such that craft can be docked into *surfer landing* to facilitate safe transfer of personnel.
Surfer Landing (or Ladder)	Docking arrangements installed on offshore facilities or vessels to facilitate access and transfer of personnel using *surfer* - type craft.
Swivel	Connecting link or device used to prevent development of twists in wire or chain cables.
Tension Control	Control facility to enable winch to be set to pull in or pay out at a specified tension.
Toolbox Talk	A meeting of the individuals due to be involved in an imminent task to review the task, individual responsibilities, equipment required, competency of the individuals, hazards, any Safe Job Analysis or Risk Assessment and or Permit to Work in place, simultaneous tasks ongoing which may affect the task and any other relevant subject.
Tow Eye / Towline Guide	Arrangement for keeping towline in centre line or midship area.
Towing Pins / Guide Pins	Device for guiding towline or pennant wire.
Towing Winch	Similar to a working winch, often geared differently. Newer towing winches have drums smaller than working winches.
Towline	Wire on towing winch used for towing.
Trigger Point	Threshold, generally relating to environmental conditions, prompting review and or risk assessment relating to the continuation or suspension of present operations.
Tug Management System	Navigation equipment on board an anchor handling vessel for an anchoring operation functioning as an interface with the installation's (MOU) main navigation equipment.
Tugger Winch	Winch provided to move items laterally on the deck of an OSV. May also be used to secure such items whilst in transit. May have remote control on newer vessels or may be controlled from the bridge on some vessels.
Tugger Wire	Steel or fibre wire used for tugger winch.

Up Weather	A position on the weather side of an offshore facility or vessel.
Weak Link	Component in any load-bearing system which is designed to fail at a pre-determined load to protect the other components in the system.
Weather Criteria	Specification of maximum allowed weather (wind, waves, etc.) when performing the operation.
Weather Side	That side of an offshore facility (or vessel) towards which the prevailing environmental forces are acting.
Weather Window	The nominated duration of specific weather criteria required to undertake a particular operation, or critical phase of same, including an allowance for any contingencies.
Working at Height	Any work undertaking where those performing it are not standing on level ground, at deck level or in other circumstances where there is a risk of injury should the worker fall (adapted from COWSP).
Working Winch	Winch for hoisting and setting anchors. Power, length, width, and diameter set the application area of the working winch.
Working Wire	Wire in working winch including termination, for example socket.

Offshore Marine
Operations

Chapter 1

Roles and Responsibilities

GENERAL RESPONSIBILITIES

All personnel. Operators, owners, and managers are required to ensure that the personnel working for them are familiar with the relevant contents of the Guidelines, as set out in the *Guidelines for Offshore Marine Operations*, published as Revision 0611-1401, dated 06 November 2013, and as subsequently amended. Whilst employers have primary responsibility for ensuring the safety of their work sites, personnel should also take care of both their own safety and that of their colleagues. Personnel must always act in such ways so as to prevent accidents and or incidents and should be empowered to "stop the job" in the event of any safety related concerns. In the interests of safety, all personnel must participate in relevant safety and working environment training activities.

Management. Management level personnel have specific responsibilities regarding the safety and performance of offshore activities. Active involvement of management is key to the delivery of satisfactory HSSE performances, combined with efficient operations. Management comprises the relevant decision makers within the framework of operating companies, logistics service providers, base operators and the owners of vessels and offshore units. It is a shared management responsibility to make available the necessary resources to ensure safe and efficient operations, including the facilitation of safe working environments and operations. This means carrying out regular visits to workplaces, with as a minimum, at least once per year. In addition, p management are expected to participate in events which promote the sharing of best practice for safe and efficient operations, and following up lessons learned from incidents and non-conformance reports to ensure any remedial measures identified have been implemented and are having the desired outcome.

Operational responsibilities. These include the minimum safety requirements that should be identified, which includes, but are not limited to, the following:

Operators and logistics companies or service providers. The role of operators, logistics companies and service providers is to: (1) establish quality assurance programmes that ensure all vessels supporting their operations are maintained and operated in accordance with agreed standards; (2) the provision of all relevant information regarding facilities which are to be supported. Typical examples of data cards used to present such information are included in *annex B*; (3) clear work specification and scope of services; (4) the assessment of consequences of simultaneous vessel operations (for example, tank cleaning versus deck cargo work); (5) the identification of hazards and acceptance criteria; (6) the development and implementation of notification formats for non-conformances, accidents and incidents; (7) the establishment of the operating company's requirements for competence, training, and certificates for the scope of work the vessel is to perform; (8) the planning for work scope follow-up; and (9) the operational manning of the vessel, as described in *chapter 4: Operational Communications and Meetings.*

Base operators. Base coordinators are required to coordinate activities between base and vessels. This means: (1) providing all relevant information regarding the base facilities which will be used, with a typical example of the data cards used to present such information are included at *annex C*; (2) the implementation of risk management processes as described in *chapter 6*; (3) provision of clear work specifications and scopes of service; (4) the risk assessment of interactions between the base and vessels; (5) competence requirements of personnel who plan, coordinate or perform loading or discharging operations; (6) provision of mechanisms and persons responsible for notifying or reporting to the operating company, authorities, etc. for non-conformances; (7) ensuring adequate and appropriate communications between base and vessels, as described in section 8 of the Guidelines *(collision risk management)*; and (8) ensure that all cargo items to be lifted from the quayside onto any vessel are visually inspected and that all potential dropped objects are removed.

Offshore facility operator. The offshore facility operator, if needed, is required to: (1) prepare a facility-specific safety zone pre-entry check list, which must be forwarded to the operator for onward transmission to the charterer; (2) provide a clear scope of work; (3) oversee the implementation of risk management processes as described in chapter 4 of the Guidelines *(operational risk management)*; (4) provision of technical system requirements needed to prevent fluid discharges from the facility (including cooling water and solids) drifting towards vessels working within the safety zone; (5) provision of mechanisms and persons responsible for notifying or reporting non-conformances, etc. to the operating company and authorities when vessels are within safety zone; (5) provision of training and competence requirements for personnel responsible for or participating in the loading, offloading and other coordinated operations with vessels; (6) ensuring that local reference systems associated with dynamic positioning arrangements used by any vessel are properly maintained; (7) planning for work scope follow-up on completion of activities; (8) ensuring adequate and appropriate communication between the offshore facility and vessels, as described in *chapter 6* of the Guidelines *(Operational Communications and Meetings).* After commencement of operations, (9) ensure vessels are advised of any subsequent changes to operational circumstances which may have an impact on the continuing work scope; and (10) ensure that all cargo items to be lifted from the facility onto any vessel are visually inspected and that all potential dropped objects are removed.

INDIVIDUAL RESPONSIBILITIES

The responsibilities of specific individuals involved in offshore marine operations are set out below.

Vessel owner / manager. Vessel owners and or managers are required to: (1) ensure that any non-conformances identified during any inspections associated with a charterer's quality assurance programme are closed out in a timely manner; (2) communicate the work scope to vessel; and (3) manage vessel operations and manning, by ensuring:

- The vessel is appropriately and competently manned and equipped for the intended work scope.
- A common working language is used on the vessel.

- An overall operational plan is prepared for all anticipated on board operations and services provided by the vessel.
- The preparation of operational conditions for vessels (i.e., defining requirements for the safe operation of vessels under all conditions, and any vessel limitations due to (for example) lack of technical redundancy.
- Incidents, accidents and safety observations are recorded, assessed and handled in accordance with an established reporting system.
- An up-to-date copy of the Guidelines is kept on board and ensuring the master, ship's officers and crew are familiar with the relevant contents.

Vessel masters. The vessel master is required to: (1) ensure that all officers, crew and other personnel on board are aware of the relevant contents of the Guidelines; (2) be, at all times, responsible for the safety of their crews, their vessel, its cargo, and the marine environment; (3) in the event of extended operations, either in port or at sea, ensure that all personnel engaged in such operations have adequate rest periods, and that effective arrangements for the transfer of responsibilities and operational awareness are implemented; (4) whilst remaining accountable at all times, delegate appropriate responsibilities to other members of the vessel's complement; (5) ensure that all onshore personnel, including representatives of the base operators, are aware of the appropriate point(s) of contact on the vessel in relation to any activities being undertaken on board; (6) approve loading plans before cargo (both bulk and deck cargo) is loaded on board the vessel; (7) review all dangerous goods declarations prior to the loading of any dangerous goods in port and offshore; (8) where relevant, refuse any cargo for which the appropriate MSDS are not provided; (9) report incidents and non-conformances; (10) inspect and approve sea fastening of cargo; (11) ensure that berth to berth passage plans are prepared for each voyage; (12) ensure all applicable field charts and relevant documents are on board and current; (13) before entering the safety zone, obtain permission from the facility manager or authorised representative for the commencement of maritime operations; (14) advise the facility of any operational limitations due to personnel, plant or environment which may have an impact on the intended work scope; (15) ensure that all cargo items to be lifted from the vessel to the facility or quayside are visually inspected and that all potential dropped objects are removed; (16) on commencement of work, advise the facility of any subsequent changes to operational capability which may have an impact on the continuing work scope; (17) when alongside an offshore facility, if extended interruption of operations occurs, decide whether to move to a safe position pending resumption of normal operations; and (18) if such a decision is made, the facility manager must be informed before moving away. The master always has the authority to stop any operation which they consider a threat to the safety of the vessel, other assets or any personnel. Other pressures must never interfere with their professional judgement. As such, the master must inform any relevant parties of conflicts of interest arising from the actions of others.

Operating or logistics company managers. Operating or logistics company managers perform overall supervision of base, vessel and installation activities. This requires them to: (1) define job performance requirements; (2) ensure that all personnel performing work on their behalf complies with the requirements of the health, safety and environment regulations; (3) manage non-conformance resolutions; (4) ensure time is allowed to perform health and safety requirements, including meetings; (5) provide up to date documentation for the master and vessel owner including necessary field charts and other relevant documentation; (6) ensure a current copy of the Guidelines is available at all locations where activities for which they are responsible are undertaken, and on all vessels supporting such operations. The operating or logistics company manager must never pressurise any master to undertake any action which, in the master's professional judgement, may compromise the vessel, other assets or any personnel.

Base managers. Base managers are required to: (1) ensure time is allowed to perform health and safety requirements, including meetings; (2) prior to loading, prepare the required documentation for the cargo to be shipped; (3) ensure that the necessary information is provided to the master in sufficient time to plan loading and discharging operations, including ensuring that dangerous goods, noxious liquids and other hazardous products are handled according to regulatory requirements; (4) ensure the master is provided with sufficient information relating to the proposed cargo so that stability calculations can be completed prior to departure; (5) ensure the master is advised of any intention to load any unusual items onto the deck of the vessel in sufficient time for any potential risks to be adequately assessed; (6) ensure the proposed stowage plan is agreed with the master, particularly when any unusual items are included in the cargo. This plan should be signed of by both parties; (7) ensure safe passage of all personnel visiting vessels, including security support; (8) arrange for all outbound cargo to be inspected prior to delivery to the vessel to ensure that it is adequately prepared for marine transportation and is free from any loose items or other potential dropped objects; (9) issue required documentation to the master for all cargo loaded on board before the vessel leaves the quayside; (10) conduct an inspection of all load carriers to ensure they are correctly certified and in proper working order before being lifted on board the vessel; (11) ensure that a cargo check list has been completed. Moreover, base managers are responsible for all HSSE compliance on the base. This means they must agree procedures to be used between all relevant parties, as well as arrange for all inbound cargo received from offshore to be adequately inspected prior to dispatch and onward carriage from the base to its eventual destination to ensure that it is adequately prepared for surface transportation and is free from any loose objects.

Facility manager. The facility manager is responsible for: (1) the safety of structure and personnel on board, and any operation within the safety zone affecting HSSE performance on the facility; (2) providing an overview of simultaneous operations; (3) ensuring operations on the facility do not present a hazard to vessels alongside, especially where over side discharges may fall onto a vessel in the immediate vicinity; (4) approving the commencement of operations and has authority to stop any operation; (5) providing proactive involvement in the risk assessment of any non-standard operations involving any vessels supporting the facility; (6) preparing the required documentation before loading is initiated for cargo to be shipped ashore by the vessel; (7) preparing documentation for transporting dangerous goods prior to loading onto the vessel; (8) submitting relevant documentation to the vessel master; (9) ensuring that the necessary information is provided to the master in sufficient time to plan loading and discharging operations, including ensuring that dangerous goods, noxious liquids, and other hazardous products are handled according to regulatory requirements; (10) ensuring the master is advised of any intention to load any unusual items onto the deck of the vessel in sufficient time for any potential risks to be adequately assessed; (11) ensuing optimal turn-around times for the performance of planned operations when vessels enter the safety zone; (12) ensuring that vessels are worked in a timely manner whilst alongside the facility so that time in close proximity to the facility is minimised. If idle, vessels should be asked to stand-by outside the safety zone; (13 issue required documentation to the master for all cargo loaded on board in a timely manner prior to the vessel departing the facility; (14) in the event of an incident or accident within the safety zone, the facility manager must inform the relevant operating company and the master of the vessel involved as soon as possible; (15) ensuring there is a good level of communication between the vessel and the facility notwithstanding all communications should take place at appropriate times and not during critical operational phases on the vessel, for example, when setting up to commence work. It is recognised the facility manager may delegate these responsibilities as required to other competent persons.

Chapter 2

Operational Risk Management

It is recognised that various terms are used throughout industry, but in this publication *Risk Assessment* (RA) shall be used as the generic term. Other commonly used terms include:

- *Safe Job Analysis* (SJA).
- *Job Safety Analysis* (JSA).
- *Task Risk Assessment* (TRA).
- *Hazard Identification Review* (HAZID).
- *Hazard and Operability Review* (HAZOP).
- *Hazard Identification and Risk Assessment* (HIRA).

Whatever terminology is used, good operational risk management is a key component to successful HSSE management. All parties involved in an operation have a duty of care to ensure the task or activity is carried out properly and safely. There are four key components to ensuring successful HSSE management. These are:

- *Risk Assessment* (RA).
- *Permit To Work* (PTW).
- *Toolbox Talk* (TBT)[1].
- *Management of Change* (MOC).

In this chapter, we will briefly examine each of these components.

RISK ASSESSMENT

The objective of risk assessment is to identify and mitigate risks to an acceptable level. If the risks cannot be mitigated to an acceptable level the work should not proceed in its present form.

Each party involved in an operation must have in place an appropriate procedure for carrying out their own risk assessments, where appropriate. Risk assessments should include all parties involved in the operations to which they relate. The risk assessment should be performed for the complete process or operation and should include relevant emergency response arrangements. Personnel performing the risk assessment must be trained and competent in doing so. Risk assessments should identify the following:

- All hazards associated with the proposed operation.
- The probability of a hazard causing harm to personnel, assets, or environment.
- The likely extent of the harm that may be caused.
- Mitigation measures.
- Assessment of the residual risk.

Associated with bullet four above are trigger points or any other changes in circumstances which may prompt the work to be stopped or the management of change process being invoked. Personnel performing tasks are required to understand the outcome of the risk assessment, including said trigger points or any other changes which would require the management of change process to be initiated. All relevant parties are responsible for ensuring the risk assessment is suitable and sufficient for their own particular tasks.

PERMIT TO WORK (PTW)

A permit-to-work system is a formalised and documented process used to control work which is identified as being potentially hazardous. It is also a means of communication between the facility, the vessel, and or base management and the personnel who are required carry out the hazardous work. The permit-to-work system used should be that adopted by the organisation in charge of the premises (or vessel) where the work is to be undertaken. As such, the permit-to-work system is to identify physical or other barrier arrangements to be put in place; be issued for a specific task, and for a specific time period not exceeding 12 hours or other clearly specified time limit; make reference to the appropriate risk assessment and its outcome; identify all lock-outs and tag-outs which should be in place before the work commences; identify restrictions or limitations in concurrent tasks; be approved and signed off by an issuing authority; confirm correct personal protective equipment is in place for the task to which the permit relates; where relevant, identify appropriate emergency response arrangements for the task to which the permit relates; and, the permit-to-work system must be effectively communicated to all parties involved.

TOOLBOX TALK (TBT)

Immediately prior to the task being carried out personnel involved in the task should carry out a toolbox talk. This should include (but is not necessarily limited to):

- Individual roles.
- Tools, methods and procedures to be used.
- Review of the risk assessment and permit-to-work.
- Promote a "Stop the Job" culture.
- Highlight all emergency actions and exit routes from the work site.
- Confirm the personal protective equipment required for the task.
- Where relevant, confirm emergency response arrangements are in place.

Personal Protective Equipment (PPE)

Personnel are legally required to be supplied with personal protective equipment appropriate to the tasks being undertaken and as identified within the procedures, risk assessments and other control measures established to ensure their health and safety. Personnel should inspect the personal protective equipment supplied for suitability and damage before use. This should be used without exception whilst the work is in progress or their supervisor advised as to why the personal protective equipment supplied is unsuitable. Examples of minimum recommended personal protective equipment requirements are shown in the table at *Annex D, Figure D-1*. Crucially, it is the individual's responsibility to use personal protective equipment correctly, look after personal protective equipment properly, and get personal protective equipment checked, maintained or replaced as appropriate.

MANAGEMENT OF CHANGE (MOC)

A management of change process should be in place for all tasks. Management of change is an important tool in preventing accidents, incidents and near misses. Tasks will normally commence and proceed in accordance with previously agreed procedures. However, should unexpected changes in circumstance occur in the course of the task the management of change process will be invoked at which time all relevant permits to work will be suspended. The task should be stopped or suspended whilst the implications of the change are reviewed. If appropriate, the risk assessment should be reviewed before resumption of the task or the toolbox talk revisited prior to the suspensions of relevant permits to work being lifted.

ACCIDENT, NEAR MISS AND OBSERVATIONS REPORTING

All accidents, incidents, near misses, non-conformances and observations are to be reported as per individual company procedures or as otherwise agreed. The objective of reporting is to establish the potential severity of the event and to ascertain whether further investigation should take place to determine immediate and root causes of the occurrence. Investigations should be comprehensive and seek to identify and implement actions to prevent recurrence. The *Root Cause Analysis* technique is a particularly powerful tool in achieving these objectives. Findings should be communicated to the parties involved and industry where relevant. All accidents or incidents within the safety zone should be reported as soon as possible to the facility and operating company managers, in addition to other statutory requirements. All accidents or incidents outside the safety zone should be reported in accordance with applicable regulations, the owner's, and other applicable procedures, and statutory requirements. As a courtesy such events should also be reported to the charterer having regard to potential reputation impact. All incidents resulting in pollution of the marine environment, including spills or releases, must be reported to appropriate regulatory authorities.

POTENTIALLY HAZARDOUS SHIPBOARD OPERATIONS

As part of their compliance with the ISM Code, vessel owners will have identified potentially hazardous operations on the vessels for which they are responsible. Owners or managers of vessels to which the ISM Code does not apply should ensure that its provisions relating to HSSE matters are complied with as fully as is practical. Typically, hazardous operations on shipboard may include, but are not limited to those listed in *Table 2-1*. Depending on the requirements of the ISM system relating to particular vessels some of these potential hazards may be grouped together, but the responsible owner should ensure that all are addressed using the risk management processes described in the earlier part of this chapter.

SIMULTANEOUS OPERATIONS

In supporting offshore marine operations vessels may be required to participate in activities involving offshore facilities or other vessels which could introduce potential hazards to personnel, equipment or the environment. Those responsible for managing such operations should ensure that the risk management processes described earlier in this chapter are complied with and that, where relevant, representatives of the respective vessel management teams are fully involved or consulted.

Table 2-1 Hazardous Operations

Nature of Hazard	Further Details
Entry into Enclosed or Confined Space(s)	Including entry into any Dangerous Spaces
General and Low Voltage Electrical Work	
(Less than 1,000 volts)	
High Voltage Electrical Work	Including work on switchboards, etc.
(1,000 volts and over)	
Hot Work	Including arc welding, cutting using gas or grinders
Work involving Critical Machinery, Operational Machinery or Control Arrangements	May include software maintenance or modifications
Work on any systems containing stored energy	Including pressurised systems or any arrangements involving rigging under tension
Work on Deck in Heavy Weather	Particularly on vessels with a low free board
Working At Height or Over Side	Working at any height where fall could result in harm to personnel
	Working outside the side rails around any open deck.
Other Non-Routine Work	Other Non-Routine Work Including, but not limited to, non-routine lifting

END NOTES

1 Toolbox Talk is commonly written as Tool Box Talk. The meaning is the same.

Chapter 3

Certification, Training, Competency and Manning

The intention of this chapter is to ensure that offshore marine operations are performed to an acceptable standard and in a controlled manner. The competence regimes in the industry are based on both international and national regulating bodies, in addition to "Best Practices" and "Guidelines". This means that while Seafarers should adhere to their Flag State requirements for their Maritime Competence, other personnel have to comply with other requirements. This chapter therefore focuses on the competencies likely to be involved on mobile units and vessels supporting offshore operations. "Competency" has been defined as the "acquisition of knowledge, skills and abilities at a level of expertise sufficient to be able to perform a task to a required standard". Alternatively, competency can be summarised using the acronym "*KATE*":

- Knowledge.
- Ability.
- Training.
- Experience.

CERTIFICATION AND COMPETENCY REQUIREMENTS

General Maritime Personnel Certification. General Maritime Personnel Certification relates to all Flag State requirements concerning the mandatory training and certification requirements in order to serve as a Seafarer; for example, Certificates of Competency, Safety Courses, Medical Certificates and similar. Compliance with the STCW[1] requirements are usually sufficient for these purposes.

Other Maritime Personnel Certification. In some instances certification relating to specialist functions may not fall within the scope of the STCW requirements and may be managed by bodies other than the vessel's Flag State. In the case of OSVs this is most relevant in the case of qualifications relating to the operators of Dynamic Positioning (DP) Systems, where the certification regime is managed by the *Nautical Institute* and some other agencies.

Functional Competence. Functional competence includes personnel having the required level of competency to support specialist functions which may not be regulated by the vessel's Flag State. Typically these would include banking cranes, rigging and slinging, winch operations, advanced first aid training on stand-by vessels, and many other similar functions. Such competency levels, which should be based on both theoretical knowledge and practical experience, relate to the vessel's function in the activities it is supporting.

Training, Experience Levels and Record Keeping. In many cases course attendance will not be sufficient to acquire those skills which also require practical experience of the work to be undertaken. Vessel owners should therefore ensure that arrangements are in place to record both the training and experience of personnel in relation to any task which may be undertaken on a facility or vessel. These arrangements should be in a form that can easily be transferred between employers if required. Individuals should ensure that such records which relate to them are updated and verified as new skills are acquired. In addition to normal operations, any training and experience should also relate to any emergency situations which might arise as the task progresses. It is the responsibility of the vessel owners or managers of facilities and vessels to ensure that personnel have the necessary competency and experience to undertake any tasks to which they may be assigned. Courses relating to any task may be arranged within the employer's organisation, including recorded "on the job" training, or by a competent external service provider. These may include crew resource management and other training courses in a simulated environment.

TEAM COMPETENCIES

To accommodate training and personal development, the competency of the overall marine team involved in a particular operation should also be considered. To promote training and sharing of expertise, less experienced personnel should be teamed up with those having a good understanding of the task being undertaken. When considering the personal development of a seafarer, owners and operators should therefore endeavour to ensure that the individual concerned can work as part of an experienced team. This is equally as important on offshore facilities as it is on vessels.

MARINE PERSONNEL COMPETENCE OVERVIEW

The STCW convention, together with its subsequent amendments, governs the majority of aspects relating to the employment of seafarers, including qualifications, hours of work, physical health and other conditions of employment. Compliance with STCW requirements should therefore be deemed sufficient for maritime personnel serving on OSVs. All vessels should be manned in order to provide all specified contracted services, unless otherwise stated in the charter party. The master must ensure that all personnel comply with current STCW Hours of Rest regulations at all times, and those relating to the *Maritime Labour Convention* (MLC). Where, for example as a result of its size, the above provisions do not relate to a particular vessel, they should be complied with as fully as may be practical. Minimum Safety and Security requirements are governed by Flag State requirements, and therefore are not addressed in the Guidelines. It is the owner or manager's responsibility to man the vessel in such a way that in all circumstances the crew are able to conduct the required operation in a safe manner. Therefore a thorough assessment of intended operations should be undertaken. Charterers should provide owners with sufficient information in a timely manner to allow these requirements to be met. The owner or manager should assess the proposed manning level in order to ensure that the level is suitable for the intended activities and, if necessary, make appropriate arrangements to ensure that all operations likely to be involved can be safely undertaken.

Operational Levels. The introduction of Operational Levels is meant to draw focus from vessel capabilities to the complexity of the operation itself. This means that competence requirements should reflect the complexity of the operations within the contract "scope of work", instead of, for example, the DP Class of the vessel. It also means that if an AHTS vessel performs cargo runs, it is the cargo run that is the essential criteria, and the competence requirement should reflect that. These Operational Levels relate **ONLY** to vessel crew. In order to set the right operational level, it is essential that the charterer specifies in the "scope of work" what kind of operations are to be expected during the contract period. It is not desirable to "up man" and "down man" during the operation. For longer term charterers the intended scope of work for the vessel, together with the process for managing any changes to the scope throughout the charter period, should normally be included as part of the charter party.

Operation Level A. Basic operations are not linked to vessel type (function / typical operations) and typically involve operations outside any safety zone; all response and rescue support operations; transits (including towing) in order to do ANY JOB, this level must be met, as this is a safe manning (SMC) requirement from the Flag State. The manning requirement is in accordance with the Safe Manning Levels of the SMC and vessel's safety management system.

Operation Level B. Standard operations involving medium complexity (function / typical operations), such as cargo operations within the safety zone, including those supported by dual-role stand-by vessels; simple low-load anchor handling operations; ROV operations outside the safety zone. Bridge manning requirements include two fully certificated STCW officers. If required, a winch operator as described in vessel's SMS. If operating on dynamic positioning bridge manning should consist of one operator certified in accordance with the vessel class notation and a second who, as a minimum, has attended the basic DP Induction course. Dependent on the qualifications and previous experience of the second operator a period of equipment familiarisation in accordance with IMCA, MTS, or equivalent recommendations may also be desirable. Engine room manning requirements include not in UMS mode (active monitoring). During these operations all machinery functions are to be actively monitored by the current watchkeeping engineer from a location adjacent to the machinery space so that, should physical intervention be required, the response time for such intervention is minimised. Deck manning is subject to risk assessment. Cargo or anchor handling operations will normally require 2 qualified seamen with appropriate operational experience.

Operation Level C. Advanced operations involving high complexity (function / typical operations), such as complex anchor handling operations, typically piggybacking, pre-laying or in deep water; close approach inter-ship operations; sub sea construction; diving support; complex ROV operations, in close proximity to surface assets; simultaneous multi-vessel operations; and vessel supported lifting operations within the safety zone. Bridge manning requirements include a minimum of two fully certificated STCW officers. If required, a winch operator as described in vessel's SMS. If operating on dynamic positioning two fully certified DPO's one of whom should be a SDPO and the other a DPO in accordance with IMCA, MTS or equivalent requirements. Engine room manning requirements include not in UMS mode. During these operations all machinery functions are to be actively monitored by the current watch-keeping engineer from a location adjacent to the machinery space so that, should physical intervention be required, the response time for such intervention is minimised. Deck manning is subject to risk assessment. Anchor handling and other operations will normally require two qualified seamen with appropriate operational experience.

Vessel Competency Matrix. The requirements of Operational Levels A, B and C in relation to bridge and engine room personnel are summarised in *Table 3-1.*

Application. The above recommendations relate to the majority of vessels supporting offshore operations. However it is recognised that in certain circumstances, particularly in relation to smaller vessels (typically less than 500 GRT) with restricted accommodation, compliance with these recommendations may not be possible. In such circumstances it is the responsibility of the vessel owner or manager to ensure that the vessel is adequately manned with appropriate procedures for the function it is required to support. Relevant Flag State and local requirements should always be complied with.

MOU Moving Operations. Due to the nature of MOU moving operations participating personnel must additionally be familiar with all aspects of such operations as follows.

Table 3-1 Vessel Competency Matrix

MANNING		OPERATIONAL LEVEL		
		A	**B**	**C**
2* Watchkeeping	Bridge	As per SMC and SMS	2 x STCW	
	Engine Room	UMS (if so classed)	Not UMS (active monitoring)	
Dynamic Positioning (if used)		1 unlimited DPO [a] or 1 restricted DPO [b]		1 unlimited DPO
Deck (marine) (excl. project personnel)		As required	Subject to risk assessment (likely to be a minimum of x 2)	

Senior watchkeepers in charge of anchor handling operations. Senior watchkeepers in charge of Anchor handling operations require relevant expertise. Watchkeepers allocated in charge of operations with no previous Anchor handling experience should perform at least five MOU moving operations accompanied by an Anchor handling experienced master, or a suitable combination of rig moves and simulator training in accordance with a training matrix and experience log, before they may command an Anchor handling assignment. Anchor handling experience gained in a chief officer role is acceptable. Masters having previous Anchor handling experience as a master or chief officer, but where this is more than 5 years ago, should have an overlap period of at least 14 days with an experienced Anchor handling master. At least one Anchor handling operation must be performed during this period.

Officers. Officers involved in Anchor handling operations also require relevant expertise. In particular officers must have a full understanding of all safety aspects of anchor-handling especially with regard to the safe use and limitations of equipment. Before participating in the Anchor handling team associated with the MOU moving operation, deck or engineer officers with no previous Anchor handling operations experience should undertake a formal offshore and Anchor handling familiarisation course or programme. This can be a combination of deck and bridge experience in a real or simulated environment. Participation in any such programme should be recorded. If supervising Anchor handling work on deck, the officer must have Anchor handling experience and be competent in Anchor handling procedures and guidelines, Anchor handling equipment set-up and function, and be familiar with the associated hazards and risks. Officers working on the bridge during Anchor handling, and may have tasks affecting the safety of those working on deck, should be familiar with Anchor handling deck work operations and the associated hazards and risks.

Vessel winch operators. Vessel winch operators should be competent in the use of winches, safety systems, functions and limitations. Ship owners should be able to document that appropriate "on the job" training or a course has been given. A training certificate should be issued by the vessel owner and or a recognised training provider.

Deck crew. Personnel assigned independent work on deck during Anchor handling operations should be familiar with the guidelines and procedures for this work, as well as general Anchor handling safety. They should also be familiar with the use of UHF and VHF radio communications. Able seamen with no previous Anchor handling experience must be suitably trained in the guidelines, procedures and safe use of equipment before assignment to independent Anchor handling work on deck. All training undertaken must to be documented. At least one member of each deck watch should have performed a minimum of five MOU moving operations.

Tow master(s). It is the responsibility of the organisation providing or employing a person to undertake the function of tow master to ensure that the individual has the competency and experience to fulfil this function. It is recommended that persons supporting this function should have participated in the moving of mobile offshore units in the following capacities: in relation to semi-submersible units acted as a stand-alone barge supervisor on such units for a minimum of three rig moves or as assistant tow master for a minimum of five rig moves; both roles should be supervised by an experienced tow master. In relation to self-elevating units, acted as a stand-alone barge supervisor on such units for a minimum of three rig moves or as an assistant tow master for a minimum of five rig moves; both roles should be supervised by an experienced tow master. Recent experience gained as master or senior watchkeeper on vessels which have been engaged in anchor handling operations of a similar nature should also be taken into account when assessing the competency of a tow master. In this context "recent experience" should be taken as being within the previous three years, though earlier experience may also be taken into account if particularly relevant. In addition, persons acting as tow master should have:

- Relevant marine knowledge and experience.
- Where necessary, appropriate qualifications, which may include STCW certification.
- Full understanding of the proposed operation, including any particular risks which might be involved.
- Appropriate knowledge of Geotechnical / Soil Conditions.
- Knowledge of Offshore Meteorology and Forecasting.
- Knowledge of DP Operations (if relevant).
- Knowledge of relevant international and local rules and regulations.
- Ability to communicate effectively in English and the local working language.

Marine representative(s). It is the responsibility of the organisation providing or employing a person to undertake the function of marine representative to ensure that the individual has the competency and experience to fulfil the function as it relates to the particular operation; the terms of reference for the role are fully understood; and the individual has been adequately briefed and has been provided with all relevant information.

Dual responsibilities and reporting functions. Situations have arisen where the same individual is acting on behalf of two or more interested parties, often involving both the tow master and marine representative functions. This should *not* be considered good practice, since the roles, responsibilities and reporting functions of the individual involved are likely to be compromised. Such situations should therefore be avoided wherever possible.

MOU winch operator. MOU winch operators should be competent in the operation of the on board winches, as well as relevant and ancillary safety systems, functions and limitations. The MOU owner should be able to document that appropriate "on the job" training has been provided.

Crane operators (including sub sea functions). Crane operators must be certified and competent in the operation and management of on board cranes, safety systems, functions and limitations. Operational experience with cranes installed on the vessel or MOU is to be logged, including operation of any heave compensation and or other particular features provided. The vessel or MOU owner should be able to document that appropriate training has been provided. For examples of training requirements refer to the *OMHEC Training Standard*[2] or local equivalent.

THIRD PARTY INDUSTRIAL PERSONNEL ON BOARD VESSELS

General. All personnel involved in the operation must be able to communicate in English or, where agreed in advance, another common work language. The vessel master is responsible for the safety of the total complement of personnel on board, the vessel and its cargo. In order to be satisfied that any third party industrial personnel on board are competent for the roles they are expected to perform, employers should ensure that appropriate documented proof is provided to the master for such verification. Contractors are required to adhere to the vessel Safety Management System, including risk management processes, as described in chapter 2 (*Operational Risk Management*).

Tank Cleaning Personnel. Tank cleaning contractors must nominate a foreman who is responsible for supervising the task. The foreman, as a minimum, must be able to speak English, or the agreed working language, and should also be able to communicate effectively with the labour under their supervision. The tank cleaning personnel must be competent in the matters described below. *Risk Assessment.* The tank cleaning foreman should be able to demonstrate to the master that they understand the principles of and are capable of undertaking a risk assessment relevant to the intended task. All employees participating in the operation must be able to understand and adhere to the outcomes of the risk assessment. *Atmosphere Testing and Tank Entry.* All tanks should be considered "dangerous spaces" which, if appropriate precautions are not taken, represent a serious risk to the personnel required to enter such compartments. The tank cleaning foreman must demonstrate to the master that they are competent and qualified to perform testing of the atmosphere in the tank to prove that it does not represent a threat to any personnel who may be required to enter that space. If the foreman is unable to competently perform atmospheric testing, a suitable chemist or other appropriate professional should be tasked with verifying the safety of the atmosphere. They must understand and know the safe and dangerous limits for oxygen, flammabilities and, if relevant, the toxicities of the former or intended contents of the tank(s). *Emergency Response and Escape.* The tank cleaning foreman must demonstrate to the master that the emergency response and escape arrangements identified in the risk assessment are in place and available if required.

DANGEROUS AND NOXIOUS LIQUID CARGOES

The carriage and handling of dangerous and noxious liquid cargoes by ship is governed by the International Maritime Organisation (IMO), and implemented by the different Flag States and Coastal States. There are no specified competence standards covering the freight of dangerous and noxious liquid cargoes on offshore supply vessels. To that end, the recommended competency levels for handling these cargoes are as set out below:

Vessel Personnel. Masters, chief engineers and certain other officers should received suitable training relating to the SOLAS[3] and MARPOL[4] requirements which include the relevant parts of the IBC Code[5] as referred to in A.673 (16) (*Guidelines for the transport and handling of limited amounts of hazardous and noxious liquid substances on OSVs*) appropriate to the vessels to which they are assigned, the IMDG Code[6] and the OSV Code[7], where relevant.

Onshore Personnel. Personnel working at the onshore base or on the offshore facility with responsibility for the declaration and shipment of dangerous or noxious liquid cargoes should receive similar training so that they have a full knowledge and understanding of the requirements that vessels must comply with when carrying such cargoes.

END NOTES

1 International Convention for Standards of Training, Certification and Watchkeeping for Seafarers (IMO Convention 1978, as subsequently amended).
2 Offshore Mechanical Handling Equipment Committee (OMHEC) Training Standard; available at https://www.hse.gov.uk/offshore/trainingstandard.pdf
3 International Convention for the Safety of Life at Sea.
4 International Convention for the Prevention of Pollution from Ships (1973/78).
5 International Bulk Chemical Code.
6 International Code for the Maritime Transport of Dangerous Goods in Packaged Form.
7 Code of Safe Practice for the Carriage of Cargoes and Persons by Offshore Supply Vessels.

Chapter 4

Operational Communications and Meetings

Offshore operations are often complex, involving many parties. Experience has demonstrated that communication failures between the various parties are often the root cause of many subsequent problems. Where possible the charterer should ensure that all parties involved come to an agreement regarding the means of communication to be used and arrangements for operational meetings appropriate to equipment available and activities to be supported. In some circumstances these arrangements may be included in the charter party. All required communications equipment is to be thoroughly tested prior to the commencement of any operation and at regular intervals whilst it is in progress. Clear and reliable communications between all parties involved are required to stop operations in the event of a dangerous situation developing. Care must be given to ensure that communications do not distract from the primary tasks in hand. All parties should be able to communicate in English and or another agreed common language. The use of dialects which may be experienced in the course of typical offshore operations is likely to lead to confusion and should therefore be avoided wherever possible. All personnel interacting with the facility and or base must be able to communicate effectively. Reference is made to chapter 1 (*Roles and Responsibilities*) and this chapter 4 (*Operational Communications and Meetings*) in relating to communications between the various parties involved.

It is crucial to maintain radio listening watches on the nominated channel in addition to appropriate emergency and calling channels. Where practical, communications between the facility deck and the vessel should be conducted on a different channel to that used for general field or control room traffic, particularly when using VHF communications. This will help to avoid confusion and enable warnings of potential dangerous situations to be communicated more quickly. If the vessel-facility communication link suffers failure or major interference, the vessel should stand off until effective communications are restored. Before operations commence, ensure there are good radio communications between the vessel and the required facility stations. During operations, facilities should avoid unnecessary communications to vessels. Personal communication to deck areas, for example, may be by UHF or VHF. Where headsets are used, any headsets worn on deck must be set at a volume which allows other sounds (waves, sea, cargo movements, warnings, etc.) to be easily heard. Due to the danger to personnel, MF and HF radio transmissions are strictly prohibited whilst alongside offshore facilities. If this is necessary, the facility manager's permission is required. If this is refused and the requirement is urgent the master must ask permission to leave the safety zone to use these frequencies. All VHF radios should be used on low power. *Radio Silence.* The facility's requests for radio silence are to be complied with in all circumstances. Vessels should ensure all conditions identified by the facility are observed. *New Technology.* As technology develops the use of new communication devices will continue to become more prolific. When introducing any new means of communication care should be taken to risk assess the implication of their use in the circumstances in which they might be employed. Currently such systems include, but may not be limited to, smart and mobile phones, email and messaging systems, video conferencing systems, and satellite communications (SATCOM) systems. *General Communications.* General communications involving vessels and offshore facilities in the course of a typical voyage are summarised in *Fig.4-1.*

OPERATIONAL MEETINGS

General meetings. Appropriate cross-party cooperation and communication is essential to safe and efficient operations. For rapid resolution of significant issues, direct communication between parties must be established through nominated individuals. The first line of offshore communication is between the vessel master and the control room, who will consult other appropriate authorities on the facility. Operating companies are responsible for establishing effective cooperation and communication between supply chain parties. All involved should participate and deliver resolutions or recommendations. The master is to keep all relevant parties informed of any issues, maintenance requirements or breakdowns which may effect the operation of the vessel. Meetings should be minuted, with minutes being forwarded to management and retained on file.

Operational meetings. Cooperation and communication between relevant parties should be regarded as a precondition for safe and efficient operations. Suggested attendees and agenda for operational meetings are summarised in *Table 4-1.*

Table 4-1 Operational Meetings

Responsible	Operator / Logistics Service Provider	
Participants	Vessel	Master, Chief Officer, Safety Delegate and others as required
	Owner	Manager, Crane Operator, Safety Delegate
	Offshore Facility	Manager, Crane Operator, Safety Delegate Operation Manager, Shipping Manager, Vessel Coordinator, Quay Foreman
	Base Operator	
	Logistics Companies	
Purpose	Team building through contact and familiarity with each others' work location and tasks	
Agenda	HSSE matters, including incident or near misses reports Variations from safe and efficient operations Feedback on measures taken following undesired incidents or non-conformances Communication Operational matters Experience transfer Improvement projects Review minutes	
Frequency	Appropriate for the duration of the operation.	

Fig.4-1 Communications with Vessels

Communications with Vessels	Parties Involved							Information Required
Essential ● Recommended, particularly for unusual or dangerous items ◐	Charter	Base Operator	Vessel Master	Area Coordinator	Offshore Installation Manager			
Voyage Phase								**Information Required**
Start of Charter	●	●	●	●				Confirmation of operating standards contact details - including telephone numbers Decision making process Particulars of all locations (including ports) Any other relevant information
Start of Outward Voyage (prior to departure)		●	●	●	◐			Voyage planning and routing anticipated weather during voyage (including potential impact on operations) Outbound / inbound cargo requirements (including dangerous goods, urgent or Special items and in-field transfers) Particular preparations at each installation (including initial backload, etc.) Potential delays and / or routing changes
Start of General Field Operations (prior to arrival in field)			●	●	◐			Any changes to routing Particular preparations at each installation (including any initial backload, etc.) Other activities in progress in field any expected delays in course of operations Anticipated weather during operations (including effect on workability at each site)
Start of Specific Operations (at each installation visited)			●		●			Confirmation of readiness to work on arrival - and to continue to completion without undue delays Shift-patterns, meal breaks, etc. operational status of all cargo handling arrangements Initial preparations necessary prior to discharge (particularly if necessary to clear deck space to receive outward cargo) Particulars of inward cargo to be loaded onto vessel (especially any dangerous goods / heavy / unusual lifts) Any items required urgently Any potential hazards in vicinity (including discharges, local obstructions, etc.) Any unusual operations during cargo operations (including fire drills, flushing, venting, etc.)
Start of Inward Voyage		●	●	●	●			Confirmation that information relating to inward cargo received by logistics service provider (including manifest, dangerous goods information, etc.) Estimated time of arrival Operations planned on arrival Vessel requirements on arrival
Completion of Voyage (or charter is appropriate)	●	●	●					Consumables remaining on board Off-hire information / survey report (if applicable)

Chapter 5

Operational Best Practice

SAFE ACCESS TO VESSELS

Masters have a primary responsibility to ensure safe means of access to the vessel for which they are responsible. This implies a "duty of care" for all personnel seeking access to or egress from the vessel. Where vessels are berthed alongside each other the Guidelines places the responsibility for ensuring safe means of access between them on the outboard vessel, but both should cooperate to ensure that personnel are able to transfer from one to the other in safety. The provision of a safe means of access to ALL vessels, whether alongside the quay or "2nd, 3rd, or more off" is of the highest importance. Failure to provide safe means of access will result in a dangerous situation with significant risk of serious injury or even death. Best practices for ensuring safe access to vessels alongside others include the same level of safety for all accessibility to vessels; including changes in height between vessels to be minimised, use of gangways and landing areas to be adequately illuminated and free of trip and slip hazards; adequately supported hand-rails or -ropes to be provided; all arrangements to be stable and adequately secured; nets provided and adequately secured; lifebuoys to be on hand in vicinity to the access point; and personnel instructed not to use any unsafe means of access.

VESSEL OPERATIONAL CAPABILITY

At all times it is the responsibility of the master to assess the risks associated with any particular activity the vessel may be requested to support. Where necessary, the chief engineer and other responsible parties must also be consulted in making such assessments. This assessment should include: (1) an appraisal of any likely degradation in the vessel's manoeuvring and station-keeping capability in the event of a failure of any safety critical system(s) or component(s), particularly in relation to the vessel's ability to safely cease cargo operations and exit the immediate vicinity of the facility should any such failure occur during the anticipated activities. Any outcomes of this assessment must be advised to the facility manager prior to the commencement of operations. Factors to be taken into account in making this assessment may include, but are not limited to, environmental criteria; (2) the thresholds and trigger points at which continuing operations will be further reviewed must be agreed with the facility manager; (3) the position that the vessel will be required to take up during proposed operation in relation to the current environmental conditions at the facility; (4) operations which will require a vessel to take up and maintain station on the up-weather side of an offshore facility will most likely involve additional risk factors which must be taken into account when undertaking the assessment; (5) the competency of the OOW to manoeuvre the vessel manually in the prevailing circumstances should this become necessary; (6) exit routes from the working location to open water clear of the facility and all adjacent structures; (7) power distribution configurations, particularly relating to vessels with diesel-electric (or similar) propulsion and manoeuvring arrangements; and (8) power utilisation of critical manoeuvring arrangements when in the vicinity of offshore facilities.

Typically, where a vessel is required to take up and maintain station close to and on the weather side of a facility the power utilisation of any manoeuvring thruster (including main propellers) should not exceed 45%. Operations in the vicinity of assets considered to be at particular risk or where ability to safely manoeuvre clear of the facility may be restricted. Such operations may include, but are not limited to:

- Requirements for vessels to maintain station adjacent to assets containing hydrocarbons which have no or minimal protection.
- Requirements for vessels to maintain station close to multiple facilities located in close proximity to each other. Typically this would include offshore facilities where additional drilling and or accommodation units have been established to support particular requirements.

Subsequent to the commencement of operations the master must continuously monitor all factors relating to the vessel's station keeping capability. Should any of these change such that the station keeping capability of the vessel changes, the facility manager should be advised without delay, particularly if bulk transfers are in progress or are planned. If, during the course of operations, the vessel is required to move from one face of an offshore facility to another the circumstances should be re-assessed taking into account the factors summarised above. If at any time circumstances change to the extent that maintaining station in the current position relative to the facility represents an unacceptable risk the current operation should be suspended forthwith and the facility manager advised accordingly, the objective being at all times to minimise the risk of contact between the vessel and the facility. Any concerns should also be communicated to the owner and the charterer's representative. Longer term concerns relating to station keeping at any offshore facility should also be communicated to its manager and also the vessel owner.

NON-ROUTINE OPERATIONS

From time to time a requirement may exist for vessels to support operations which, by their nature, may be unusual or outside the range of activities normally supported. The Guidelines do not advocate that such operations should be curtailed or restricted, but instead seek to draw attention to the additional risks which may be involved and to recommend that, when proposed, appropriate specific task-based risk assessments, as described in chapter 2 (*Operational Risk Management*) are undertaken by the personnel involved. Operations which may be considered non-routine include, but are not limited to, the following.

Weather side working. Reference should be made to section 8.11 (*Weather Side Working*) of the Guidelines for guidance relating to procedures if requested to take up station on the up-weather side of an offshore facility. If supporting operations at offshore complexes consisting of several structures located in close proximity to each other, which may or may not be linked by bridges and may also include mobile offshore units, masters should be conscious of potential "drift on" situations developing in relation to platforms or other units apart from that at which the vessel is presently located.

Certain lifting operations. Certain lifting operations involving the transfer of cargo between a vessel and an offshore facility should not be considered as routine but should be the subject of a separate specific risk assessment. These include, but are not limited to, operations:

- Requiring the use of a crane's main block.
- Involving the lifting of long cargo items, particularly where it is necessary to use two stinger pennants from the crane's hook.
- Which require personnel on the vessel to connect or release lifting rigging using any means other than safety hooks.
- Involving the lifting of cargo items where rigging has not been pre-installed.

SOFTWARE MANAGEMENT AND MAINTENANCE

The operation of a wide variety of equipment, some of which may be safety critical, on modern facilities (including vessels) is dependent on software based control arrangements. It is therefore essential that the management and maintenance of all such control arrangements is subject to the same rigour as any other critical system installed on the facility or vessel. Any subsequent changes or updates should then be controlled and recorded in the Preventative Maintenance System (PMS) as they occur in order that a full audit trail of such amendments can be maintained, as happens in the case of modifications or repairs to other equipment.

DYNAMIC POSITIONING ARRANGEMENTS

General requirements. Any vessel chartered and approved to maintain station by means of dynamic positioning within the safety zone around any offshore facility, should observe and comply with the guidelines published by the IMO and supplemented by further guidance issued by IMCA, MTS or other similar trade associations, as updated from time to time. It is the responsibility of any vessel owner responsible for operating DP vessels within the safety zone of any offshore facility to ensure that these requirements are understood and complied with.

Locations, care and maintenance of local reference systems. It is the responsibility of the "owner" of any local, radar or optically based reference system used to support vessels maintaining station by means of dynamic positioning to ensure that it is correctly sited on the facility and that suitable arrangements have been established for its care and maintenance. Where any component of a reference system which forms part of a vessel's inventory is passed to an offshore facility to support operations at that location a document package including information regarding preferred location of the component and its care and maintenance should be transferred at the same time. Where practical, reflectors used with optically based reference systems should be sited clear of commonly used walkways or decks where containers are stored since the presence of retro-reflective material on cargo items or personal protective equipment may result in false signals being returned to the sensor arrangements on the vessel.

Optical reference systems, environmental degradation. In fog, mist, falling snow, heavy rainstorms or other conditions similarly restricting visibility the performance of optically based systems may be seriously degraded. Depending on wind direction discharges from the facility may have a similar effect. If selected as one of the position reference systems for a vessel maintaining station by means of dynamic positioning the personnel responsible for monitoring and operation of these arrangements should be aware of the potential for their degradation in such circumstances.

SIMULTANEOUS OPERATIONS (SIMOPS)

Simultaneous operations in this context refer to circumstances where two or more vessels are supporting activities within a facility's safety zone at the same time or operating elsewhere in circumstances whereby actions undertaken by one may have affect the other(s). Any hazards likely to arise during such operations should be addressed using the risk management process, as described in chapter 2 *(Operational Risk Management)*.

TOWING OPERATIONS

For guidance relating to towing operations, refer to chapter 9 *(Anchor Handling and MOU Moving)* and chapter 10 *(Project Support Operations)* for further information.

DISCHARGE FROM FACILITIES

Vessel masters must cease operations and move clear of the facility if at any time there is any concern whatsoever that discharges from any facility are posing a threat to the well being of any personnel on the vessel, affecting visibility or compromising the performance of optical reference systems. Any such concerns must be reported immediately to the facility manager where it should be followed up as a matter of urgency. Some facilities may be fitted with "auto-dump" or "auto-vent" arrangements designed to automatically empty tanks to sea or purge pressurised systems to atmosphere if certain threshold values are exceeded. Wherever practical such arrangements must be disabled whenever a vessel is approaching or working alongside the facility, and their status advised to the vessel as part of the pre-operational checks. Where this is not practical, the status of all relevant systems must be checked by the facility prior to giving the vessel permission to enter the safety zone to assess the likelihood and consequences of such an event occurring. The vessel must be advised of the outcome of this check and the master, at their sole discretion, will decide whether the facility can be safely supported in the prevailing circumstances. These arrangements should continue to be checked at frequent intervals whilst the vessel remains alongside.

OFFSHORE TRANSFER OF PERSONNEL TO AND FROM THE VESSEL

Requirements. Circumstances may arise where it is necessary to transfer personnel to or from a vessel whilst it is offshore. These may include requirements for personnel to be moved between an offshore facility and the vessel involved, or between it and another in the vicinity. The preferred means of effecting such transfers will normally be by helicopter or, where conditions are suitable, by specialised small craft subject to the facilities and or vessels involved being suitably equipped and personnel having had the correct training. Alternatively, where the vessel is providing accommodation support in close proximity to an offshore facility a gangway or bridge link between the two will normally be provided. Such transfer methods will be the subject of specific risk assessments and particular requirements, precautions, procedures and, where appropriate, combined operations safety cases will have been developed. These are therefore seen as being planned activities, consideration of which is outside the scope of the Guidelines. A requirement to transfer personnel may arise, however, when the methods described above are not available, necessitating the use of other arrangements. The equipment used for this purpose may include:

- Transfer baskets or other forms of carrier lifted by a crane on the facility.
- Other small craft where no such arrangements exist.

The remainder of this sub-section relates to the preparations required and procedures to be observed when using such equipment.

General Preparations, Precautions and Procedures to be Observed

Risk management. The risk management process, as described in chapter 2 *(Operational Risk Management)* should be complied with whenever transfers of personnel are being contemplated. In some instances transfers by means other than helicopter may take place on a regular basis, being considered the safest or most practical means of moving personnel from one location to another. Typically, such operations will involve the use of small craft specifically designed for the purpose to move personnel between offshore facilities and or vessels, all of which have been provided with docking arrangements designed and constructed for that purpose and compatible with those on the craft in use. Typical examples are the "surfer" ladders in use in many benign areas of operations, where the bow of the craft is engaged into the guides of the landing which hold it in place allowing personnel to safely step from one to the other. Similar arrangements are utilised on many small offshore structures, including wind turbines. In such circumstances, whilst the full risk management process should be complied with prior to the commencement of operations, it should not be necessary to undertake this exercise before each transfer. However, arrangements should be in place to ensure that prior to each transfer the personnel are properly briefed as to the precautions to be observed. Furthermore, the original risk assessment should be reviewed at frequent intervals to ensure that the outcomes remain valid. If, for any reason, this is no longer the case the entire process should be repeated. Where other equipment or arrangements are proposed, including the use of lifted transfer baskets or the use of small craft not specifically designed for the purpose it is unlikely that such a generic approach will be acceptable. The full risk management process may therefore be required before each operation, though a series of transfers involving the same equipment and principal personnel may be considered as a single operation.

Authorisation for personnel transfers. The personnel transfers described in this section should be the subject of approval by the persons in charge of the offshore facilities and or the vessel(s) involved. Where transfers by means other than helicopter take place on a regular basis and are considered the safest or most practical means of moving personnel from one location to another authorisation for each such activity is unlikely to be required. However, as described above, the original risk assessment should be reviewed at frequent intervals to ensure that the outcomes remain valid. If, for any reason, it is considered prudent to repeat the entire risk management process further transfers using the method involved should be the subject of renewed authorisation. Where other equipment or arrangements are proposed it is unlikely that such a generic approach will be acceptable. Each operation should be individually authorised, though a series of transfers involving the same equipment and principal personnel may be considered as a single operation.

Consent for transfer. Personnel requested to transfer between offshore facilities and or vessels by the methods described in this section should be made aware of the risks involved, together with precautions and procedures to be observed. On having received the relevant briefing personnel should positively indicate their willingness to be transferred by means of the method proposed, or, alternatively, refuse without sanction.

Suitability of equipment. All equipment utilised to transfer personnel between offshore facilities and or vessels by the methods described in this section should be fully fit for purpose and in compliance with the regulations of the jurisdiction in which the operation takes place. Further recommendations relating to specific items of equipment are included in the relevant sections below.

Storage and maintenance of equipment. All equipment utilised to transfer personnel between offshore facilities and or vessels by the methods described in this section should maintained and stored in accordance with the original equipment manufacturer's (OEM) instructions.

Experience and competency of supervisors and operators. Overseeing supervisors and operators of equipment involved in the transfer personnel between offshore facilities and or vessels by the methods described in this section should have previous experience of the operations involved and be assessed as competent to undertake the tasks assigned to them. This includes, but is not limited to the following functions:

- Supervisors of operations.
- Crane Drivers, where transfer is by basket or carrier.
- Coxswains, where transfer is by small craft.
- Attendant personnel, including deck or craft crews.

Access to and egress from transfer areas. Access and egress routes to or from the transfer area on the offshore facility or vessel should be clearly marked, dry, and clear of all obstructions or trip hazards. Where necessary, a non-slip coating should be applied to steel decks or other alternative arrangements put in place.

Communications. The means of communication between the various personnel involved in the transfer operations will have been identified during the risk management process. All such means of communication should be in place and their correct operation verified prior to the commencement of any transfer activities.

Clear View of Transfer Areas. Wherever possible personnel supervising the activities described in this section should have a clear view of all phases of the entire transfer operation. Further recommendations relating to specific transfer methods are included in the relevant sections below.

Capacity of Basket, Carrier or Craft. The capacity of any basket, carrier or craft used in the course of the activities described in this section will be determined by the original equipment manufacturer of the equipment. This capacity should not be exceeded at any time.

Personal Protective Equipment and Effects. Personnel being transferred by any of the methods described in this section should be provided with appropriate personal protective equipment. Dependent on the area where the transfer takes place such equipment may include a watertight immersion suit, thermal protection, life jacket or buoyancy aid[1]. Personal Locator Beacons, where detection and tracking facilities available Personnel should be given a briefing regarding the correct donning and use of the equipment. Before boarding the basket, carrier or craft the PLB should be checked by the person supervising the transfer. Personnel should not wear any clothing or carry any items which could restrict their mobility or interfere with the correct operation of any protective equipment. In some cases a small quantity of personal effects may be included with the transfer of personnel. However, this will involve additional space and or weight requirements which should be taken into account when assessing the available capacity of the basket, carrier or craft. If carried, such effects should be stowed and secured in such a manner that escape routes are not obstructed. Where the simultaneous carriage of personnel and their effects would compromise the capacity of or obstruct escape from the basket, carrier or craft arrangements should be made for each to be transferred separately. In general, the policies, practices and equipment relating to the transportation of personnel by helicopter are also relevant to the transfers described in this section.

Compliance with Supervisor's Directions. Personnel being transferred by any of the methods described in this section should comply with the directions of the supervisor overseeing the operation.

Availability of Rescue Facilities. Whilst personnel are being transferred by any of the methods described in this section suitable rescue facilities should be available at immediate notice. Where a stand-by vessel is in attendance, if not directly involved in the transfer operation, its master should be advised and requested to bring his rescue facilities to an immediate state of readiness. If no such vessel is in attendance, or is itself involved in the transfer operation, alternative arrangements, which may involve fast rescue boats or craft installed on other vessels, should be identified and agreed before the persons in charge give the necessary authorisation.

Environmental Restrictions. The transfer of personnel by the methods described in this section should not be undertaken where the environmental conditions were such that increased risk would be incurred. Typically, such operations should not proceed where the prevailing conditions include one or more of the following:

- Wind speeds in excess of 20 knots (10 metres / second) at a height of 10 metres (32 feet) above sea level.
- Significant wave heights in excess of 2.5 metres (8.2 feet).
- Horizontal visibility of less than 500 metres (1,640 feet).
- Heavy accumulations of snow or ice on landing areas, access and egress routes, etc.

Further restrictions relating to specific transfer methods are included in the relevant sections below. Furthermore, these operations should not normally take place in hours of darkness. Where this is deemed essential by the relevant persons in charge additional precautions are likely to be required, which may include, but are not limited to, the following:

- Ensuring that illumination of all transfer areas in adequate.
- Ensuring that life jackets or buoyancy aids are fitted with high intensity strobe lights.
- Ensuring that retro-reflective tape on overalls or immersion suits is not obscured.

Transfer operations undertaken outside environmental limits or in the hours of darkness should be the subject of a full risk assessment process and specifically authorised by the persons in charge on the relevant offshore facility and or vessel(s).

Record Keeping. The persons in charge on the offshore facility and or vessel(s) should ensure that full particulars of any transfers as described in this section are recorded in the relevant log-books and that the register of personnel on board the facility or vessel(s) is revised as soon as possible.

Particular Preparations, Precautions and Procedures. Recommendations relating specific preparations required together with the precautions and procedures to be observed relating to each of the means for effecting personnel transfers described in this section are as follows.

Use of Transfer Baskets or Carriers lifted by Facility Crane. This sub-section relates to the use of baskets or other carriers lifted by the cranes on an offshore facility to transfer personnel between it and a vessel close alongside. Recommendations which should be observed include any cranes to be used for this purpose should comply with rules or codes in force within the jurisdiction where the operation will be undertaken. These may vary from area to area but particular attention should be paid to hoisting and braking arrangements. Baskets or carriers to be used for this purpose should also comply with the rules or codes in force within the jurisdiction where the operation will be undertaken. All equipment to be used for this purpose should be thoroughly inspected by competent persons at periodic intervals, as required by the rules or codes of the jurisdiction within which they will be used. In general, baskets or carriers incorporating a rigid frame which provides protection for occupants are preferable. Baskets or carriers which do not incorporate this feature may only be acceptable for emergency use in some jurisdictions. Baskets or carriers should be rigged or otherwise fitted out in accordance with manufacturer's instructions. A basket or carrier should be fitted with sufficient buoyancy to support the unit itself and its occupants in the event of entering the water. Buoyancy should be distributed to prevent inversion should such an event occur. Baskets or carriers should be visually inspected by a competent person before each operation to ensure that all rigging, fixtures and fittings remain fit for purpose and secure. Clear lift-off and landing areas should be identified on facility and vessel. Such areas should as a minimum within a radius from centre of 1.5 x basket diameter be free of obstructions or trip hazards. Outside the lift-off / landing area there should be no obstructions extending more than four metres (13 feet) above the deck within eight metres (26 feet) of its centre and beyond this, within a distance of 20 metres (65 feet) from the centre within an arc of 180 degrees. Appropriately briefed personnel should be in attendance for both lift-off and landing to assist in controlling the movement of the basket or carrier at these critical phases of the operation. In particular, such personnel should be briefed in the use of the attached tag lines.

Any other work in the vicinity of the lift-off and landing areas should be suspended whilst the transfer is in progress. In addition to the environmental restrictions referred to above, transfers of personnel using baskets or carriers should not proceed when the prevailing conditions include:

- Vertical visibility of less that 100 metres (328 feet).
- Air temperature of -10°C (14°F), particularly if wind is also present.

Prior to the commencement of the transfer the master should confirm that the vessel is stationary and that its station keeping arrangements are fully operational. Throughout the course of the transfer the crane driver should have a clear and unobstructed view or the carrier or basket and its occupants. If, for any reason, this is not possible an experienced banksman should direct the crane driver. The banksman should be clearly identified and visible to the crane driver at all times. The route of the transfer should be planned so that the basket or carrier is always well clear of any exhausts, discharges or obstructions. After the basket or carrier is lifted from the deck of the facility the crane should be slewed so that it is over the water, whereupon it is lowered to a height of approximately two metres above the vessel's cargo rail. The basket or carrier should then be moved to a position over the designated landing area on the vessel before being finally lowered onto its deck. Transfers from the vessel to the installation should follow the reverse route. The basket or carrier should always be lowered with the hoisting mechanism engaged. Free-fall or non-powered lowering should not be used except where the hoisting mechanism fails whilst the basket or carrier is occupied. If considered necessary a person experienced in this method of transfer may accompany other personnel who may be less familiar with it. A small quantity of personal effects can be carried in some types of baskets or carriers, but not in others. If carried, such items should be stowed and secured in such a way that escape routes from the basket or carrier are not obstructed. Personnel to be transferred should only approach and board the basket when instructed by the supervisor. On boarding, personnel should secure themselves in the basket or carrier as instructed during the preparatory briefing. On landing on the deck of the facility or vessel personnel should release themselves and disembark the basket only when directed by the supervisor. They should then clear the immediate area using the route indicated. Personnel not directly involved in the transfer should remain in a safe haven well clear of the operation, except as otherwise directed by the supervisor.

Use of Small Craft. This sub-section relates primarily to the use of other small craft deployed from a larger host vessel to transfer of personnel between vessels. Such craft may typically include the following:

- Fast rescue boats mobilised on vessels in compliance with SOLAS requirements.
- Fast rescue craft or daughter craft mobilised on stand-by vessels.
- Small work-boats mobilised on a variety of vessels.

Recommendations which should be observed include personnel transfers involving only vessels should not take place within the safety zone around any offshore facility. The relevant facility management teams should be advised of the intention to undertake any such transfer, together with the masters of any attendant response and rescue vessel, if itself is not directly involved in the operation. Any craft to be used for this purpose should comply with rules or codes of the host vessel's Flag State or those of the jurisdiction where the operation will be undertaken. Any craft to be used for this purpose should be thoroughly inspected by competent persons at periodic intervals, as required by rules or codes of the host vessel's Flag State or those of the jurisdiction where the operation will be undertaken. Any craft used for this purpose should be constructed with a rigid or semi-rigid hull. Fully inflatable craft are not normally acceptable for this purpose. If permanent fendering or similar arrangements are not incorporated into the hull design suitable portable fenders should be provided. Sufficient buoyancy to support the craft itself and its occupants in the event of swamping should be installed. Craft fitted with self-righting arrangements are to be preferred. If practical, where the principal propulsion consists of a single engine and drive train an auxiliary system should be provided, for use should the principal arrangements fail. Where the vessels involved are equipped with identical craft, with the same means of deployment and recovery being installed on both, "davit to davit" transfers are to be preferred. Where fitted, permanent rigid ladders should be used, subject to their being in good condition. Typically, such arrangements are fitted on cargo barges and similar units. Where such ladders are not fitted or are in poor condition portable ladders may be provided. Portable ladders supplied for this purpose should comply with IMPA[2] requirements. Stanchions, handholds and other arrangements to facilitate the safe transit of personnel from the ladder to the deck of the vessel and vice versa should comply with IMPA requirements. Personnel to be transferred should only board the craft when instructed by the supervisor. On boarding, personnel should take their seats and secure themselves as instructed during the preparatory briefing or as directed by the Coxswain When in transit personnel being transferred should remain seated or move around with caution. On arrival at facility or vessel personnel should disembark the craft only when directed by the Coxswain. They should then follow the directions of the supervisor. Personnel not directly involved in the transfer should remain clear of the operation, except as otherwise directed by the supervisor. Whilst the recommendations above relate principally to small craft deployed from a larger host vessel they may also be appropriate for other craft capable of autonomous operation.

Further Guidance

Further guidance relating to the transfer of personnel between offshore facilities and or vessels may be found in the documents listed in *Table 5-1*.

Table 5-1 Further Guidance

Source	Document Particulars	
	Number (if known)	**Title**
IMCA	M202	Transfer of Personnel to and from Offshore Vessels

SECURITY

The vessel and or facility is to comply with the ISPS Code[3] where there is a requirement and any additional coastal or flag state requirements.

OPERATIONS IN ENVIRONMENTALLY EXTREME CONDITIONS

Guidance on operations in environmentally extreme conditions is included at *Annex E.*

END NOTES

1 Inflatable life jackets or buoyancy aids are normally to be preferred. Inherently buoyant marine life jackets provided to comply with SOLAS requirements are bulky and likely to obstruct movement.
2 International Maritime Pilots' Association
3 International Ship and Port Security Code

Chapter 6

Collision Risk Management

Vessel owners and masters should ensure that any operations which involve approaching, working alongside and departing from any offshore facility are at all times undertaken in accordance with the best practices described below.

SAFETY ZONES

Most offshore facilities will be protected by the establishment of a safety zone around the structure, unit or vessel. The best practices set out in the Guidelines, and subsequently this publication, have been developed on the presumption that such a safety zone exists but it should be noted that some offshore facilities, particularly vessels, may not be protected by such a zone. That said, it is strongly recommended that when attendant vessels are approaching any offshore facility the practices described in this chapter should be observed, irrespective of whether a safety zone has been established around the facility.

BRIDGE TEAM ORGANISATION AND MANAGEMENT

It is the responsibility of the vessel owner and master to ensure that the team directing operations from the bridge have the necessary experience for the proposed operations such that all activities can be undertaken in a safe and expeditious manner. Matters which may require particular consideration include, but are not necessarily limited to competencies, distractions, situational cognisance, awareness of environmental conditions, hand over procedures, and precautions against the onset of fatigue.

Competencies. At any time competencies of personnel available within the bridge team should comply with those identified in the relevant operational level for the current activities, as described in chapter 3 *(Certification, Training, Competency and Manning)*.

Distractions. Each member of the bridge team should be able to concentrate on their primary responsibilities. Other activities should only be undertaken when they will not compromise such responsibilities. Any members of the team who find themselves in a situation where primary responsibilities are being compromised by additional activities should immediately cease such activities, drawing this to the attention of the senior watchkeeper.

Situational Cognisance. Typically, modern marine equipment installations include a variety of aids to provide bridge team members with the navigational information necessary for the safe operation of the vessel. However, maintenance of a visual watch at all times remains an important part of the bridge team's responsibilities and should not be overlooked.

Awareness of Environmental Conditions. The bridge team should use all means at their disposal to ensure that they remain aware of prevailing environmental conditions. They should also be aware of any "trigger points" which have been identified in relation to any operations presently being undertaken. In the event of environmental conditions changing such that the threshold levels in "trigger points" are (or are likely to be imminently) exceeded the bridge team should assess whether current operations can continue or should be suspended until circumstances improve.

Hand Overs. Adequate arrangements should be in place to ensure that at the change of each watch member of the bridge team is able to give their relief a complete briefing regarding the status of present activities and the vessel's current operational status. In some circumstances, where complex operations are being undertaken, clear bridge team relief procedures should be in place to ensure a positive hand-over. Consideration may be given to arranging for members of the bridge team to be relieved at different times to ensure continuity of awareness within the team. Requirements for written records of hand-overs, to be signed off by all watchkeepers, may exist for some circumstances which should be described in the vessel's SMS Manual.

Precautions against the onset of Fatigue. In all but the most extraordinary circumstances international legislation relating to hours of work and rest periods should be complied with. Certain operations may require an unusually high level of control for extended periods. Personnel involved in such operations are therefore required to maintain an unusually high level of concentration with the result that early onset of fatigue is likely. In such circumstances arrangements should be made for the relevant personnel to be relieved more frequently than might be normal practice. Operations where such arrangements might be prudent should be identified at the early planning stage and appropriate measures put in place at that time. Operations likely to fall into this category should be risk assessed to ensure that the provisions of the *Manila Amendments* to the STCW Convention, 2010, are fully adhered to.

APPROACHING LOCATION

Wherever practical, when approaching any facility vessels should set a course which is off-set from it, at a tangent to the safety zone. This course should take the vessel to a position where it can be set up for intended operations and the check lists completed in a drift-off situation.

SELECTION OF STATION KEEPING METHOD

Following an assessment of the operations to be supported together with the prevailing and forecast conditions, the most appropriate method of station keeping whilst in the vicinity of the offshore facility will be selected at the discretion of the master or senior watchkeeper on duty at the time. The choice of station keeping method should be advised to the facility as part of the pre-entry process.

PRE-ENTRY CHECK LISTS

Prior to entering the safety zone at any facility the pre-entry check list for the vessel should be completed. Completion of these check lists should be viewed as a safety-critical function. A typical example of such a check list is included at *Annex F*. Each check list should be signed off by

ALL watchkeepers. Copies should be retained on file for audit for a limited period, circa three months. Where laminated check lists are in use an entry should be made in the vessel's log of each such use, together with a summary of the outcomes. Electronic copies of signed-off check lists will be acceptable and should be filed in a suitable manner.

CHANGE OF CONTROL STATION OR OPERATING MODE

Whenever control of a vessel is transferred to another station or a different operating mode is selected it should be ensured that all manoeuvring arrangements are responding as anticipated prior to undertaking any operations in the close proximity of an offshore facility, another vessel or other obstruction.

SETTING UP BEFORE MOVING ALONGSIDE

Vessels should set up in the vicinity of the face to be worked on the appropriate heading at a distance from the facility of not less than 1.5 ship's lengths in a drift-off situation or 2.5 ship's lengths in drift-on circumstances. When setting up to work in a drift-on situation the vessel should not be directly up-weather and or up-tide of the facility. The set-up position should also take into account any obstructions in the vicinity of the intended working location. Prior to moving from the setting up to the working location sufficient time should be allowed to ensure that all station keeping arrangements are stable and environmental factors can be fully assessed. It is suggested that a minimum of 10~15 minutes is allowed for this or as otherwise required by the vessel's operating procedures.

USE OF DYNAMIC POSITIONING

Section 8.7 of the Guidelines *(Setting Up before Moving Alongside)* relates to all vessels, and further requirements may relate to those maintaining station by means of dynamic positioning. Vessel specific directions and guidance relating to the use of dynamic positioning facilities for station keeping will be included in operating procedures prepared by the equipment manufacturer and or owner. These should be complied with at all times. Further guidance is included in chapter 5 *(Operational Best Practice)*.

IN OPERATING POSITION

Whilst alongside the facility, power consumption, thruster utilisation and environmental factors must be monitored on a regular basis, particularly if working on a weather face. Similarly, actions required to depart from the facility at short notice, should this be necessary, should be continuously reviewed. The exit route to depart from the immediate vicinity of the facility should be reviewed at the same time. If, for any reason, there is any concern regarding the vessel's ability to maintain position operations must be suspended and the vessel manoeuvred to a safe position clear of the facility. Such action is at the sole discretion of the master or senior watchkeeper. Further guidance is included in chapter 5 *(Operational Best Practice)*.

CHANGE OF OPERATING LOCATION

Where it is necessary for the vessel to move from one working location to another such movement should be carefully planned and executed. Wherever practical, risks associated in moving between locations should be assessed and personnel instructed accordingly. Wherever practical, if moving from one working face to another the vessel should avoid passing up-wind and or up-current of the facility. It should move well clear of the facility, move to the appropriate setting up location and carry out the setting up procedure described above prior to moving into the new working position. The facility should be kept fully advised regarding the progress of any move between working locations. If available, a consequence analyser may be used in simulation mode as an aid in assessing the implications of moving from one working location to another, However, the availability of this aid should never be considered as a substitute for the proper planning and implementation of such a move so that it is executed in a safe and controlled manner.

WEATHER SIDE WORKING

Any potential requirements to work on the weather side of a facility must be risk assessed as described in chapter 2 *(Operational Risk Management)* prior to moving into the set-up position and continuously thereafter until the relevant operations have been completed. When preparing to work a weather face, the vessel must not set up directly to windward of the facility, but in a drift off position so that in the event of a power failure whilst setting up the vessel will drift clear of the facility. Where, at any location, tidal or other currents are significant, similar precautions should be observed.

REQUESTS TO STAND BY FOR FURTHER5 INSTRUCTIONS

The risk of contact between an offshore facility and a vessel operating are increased if the two remain in close proximity for extended periods. If, therefore, for any reason operations at a facility cannot be completed and a vessel is requested to stand by for further instructions, cargo, etc. it should move to a location at a safe distance from the facility and in a drift off position. When returning to an operating location the pre-entry checks and set-up procedures as described above should be repeated.

EXTENDED AND PROTRACTED CARGO HANDLING OPERATIONS

The potential risk of contact between any vessel and facility is reduced when the time that the vessel is in close proximity to the facility is minimised. It is the expectation that the facility personnel will plan operations to minimise this time alongside, but should the master believe that this is not the case, resulting in the vessel having to remain alongside for protracted periods this should be brought to the attention of the PIC of operations at the facility. Where performance reporting arrangements have been made by the charterer such events should also be reported through this channel.

DEPARTMENT AND COMMENCEMENT OF PASSAGE

In all cases a safe exit route should be selected, taking the vessel well clear of all hazards, including any other vessels and to leeward of the facility. In all cases changes in operating mode from position keeping to passage making should not take place within 1.5 ship's lengths of the facility if departing from the lee side, or within 2.5 ship's lengths if departing from the weather side. Furthermore, if departing from the weather side such changes in operating mode must only be implemented in a drift off position.

FIELD TRANSITS

Some offshore developments may consist of a number of independent facilities. In some instances vessels which are not supporting or undertaking operations within the safety zones around such facilities may be required to pass through the development. When making such a field transit courses should be planned so that, where practical, the vessel passes at a distance of at least one nautical mile (1,852 metres; 6,076 feet) from each facility and any operations which might be in progress in its immediate vicinity.

OTHER RECOMMENDATIONS TO MINIMISE COLLISION RISK

Other recommendations to minimise the risk of contact between offshore facilities and their attendant vessels are included throughout the remaining chapters of this publication. These do not appear in this chapter since it is considered they are more appropriate in the general context of the subjects in which they are included.

Chapter 7

Logistics and Cargo Handling Operations

This chapter includes guidance on best practice for logistics and cargo handling operations which should be complied with by all the relevant parties involved in the course of a typical voyage to and from the offshore destination(s). When planning to load any cargo, on or under deck, on an OSV it is the joint responsibility of charterer, owner, master and base operator to ensure that the proposed vessel is fully fit for purpose and in compliance with all relevant requirements relating to the safe carriage of the goods or products concerned. Compliance with relevant international legislation, together with the rules or codes of the vessel's Flag State and those of the regional authorities in its present area of operations is included in this requirement. The charterer, owner, base operator and master should ensure that all personnel who may be involved in the loading or discharge of cargo are appropriately qualified and competent in the handling and carriage of the goods or products involved. This requirement also extends to other personnel who may be mobilised to provide any support services which might be necessary, including surveyors and other quality assurance specialists. Whilst these responsibilities relate particularly to the carriage of dangerous goods and inflammable, noxious or otherwise hazardous liquid products they also relate to all other cargoes carried on offshore supply vessels.

NOTIFICATION OF ANY UNUSUAL CARGO ITEMS

Where there is an intention to ship any unusual cargo items on an offshore supply vessel, the base operator should advise the relevant master in a timely manner in order that any risks associated with the shipment can be properly assessed and appropriate preparations made. Items falling into this category are referred to in sections 7.3.2 *(Certain Lifting Operations)* and 9.13 *(Unusual Cargo Items Loaded onto Vessel Decks)* of the Guidelines.

DECK SPACE MANAGEMENT AND BACK LOAD CARGOES

Congestion on the cargo decks of both vessel and offshore facility can result in the development of situations hazardous to personnel or equipment. Except where rigorous planning of logistics support is in place or where previously agreed and confirmed in sailing instructions it is considered good practice for a vessel to arrive at an offshore facility with approximately 10% of its usable deck clear and ready to receive initial back-load. This allows sufficient space to be cleared on the facility's deck before any cargo is taken up from the vessel. Wherever possible, this clear deck space should be contiguous. Subject to discussion with the master this recommendation may be waived at the last facility at which cargo is back-loaded onto the vessel prior to its return to base when all deck space may be utilised, but only on the understanding that it will not subsequently be diverted to support another offshore location on its inward voyage.

CARGO PLANS

In the course of the initial loading at its shore base(s) the master should ensure that a record of the cargo loaded on board is maintained. This should show the locations of the "blocks" of cargo for each facility to be supported during the forthcoming voyage, together with number of lifts in each block and other relevant details. Locations of any unusual cargo items should be clearly indicated. The cargo plan may be further supported by photographs of the vessel's deck. There is normally a requirement for this plan to be forwarded to the base operator on completion of loading, who will subsequently arrange for it to be forwarded to the facilities to be supported in the course of the subsequent voyage. The plan should be updated as the voyage progresses. A typical deck plan is illustrated at *Annex G*. Other examples, based on electronic software packages exist, and may be more easily transmitted through the communications channels in use. A table or drawing showing the contents of the vessel's under-deck cargo tanks should also be prepared and forwarded to the base operator as described above.

SAILING INSTRUCTIONS

Prior to a vessel being dispatched on any voyage delivering cargoes to one or more offshore facilities the base operator or logistics service provider, in conjunction with the charterer, should furnish it with a comprehensive set of sailing instructions. These instructions may include, but are not limited to, the following:

- The cargo manifest (which includes details of items loaded on the vessel).
- Any specific information regarding cargoes on board, including Material Safety Data Sheets (MSDS) and particular hazards associated with any cargoes.
- Any particular precautions relating to the care of any cargo.
- Passage planning (routing) for the voyage.
- Facility data cards, if not already held on board.
- Reporting requirements.
- Any changes in contact details.
- Any other special instructions or relevant information.

WEATHER FORECASTING

Arrangements should be made with a reputable weather forecasting service provider, experienced in the preparation of offshore forecasts, to prepare and promulgate weather forecasts extending, where practicable, up to five days (i.e., 120 hours) for the relevant locations. Such forecasts will generally be arranged by the charterer and should be made available to masters of all vessels operating on their behalf. The weather forecasting service provider may also be able to prepare more specialised information on request, including longer term forecasts, met-ocean statistical analyses, etc. if this is required for any particular purpose. It is also the master's responsibility to ensure that forecasts from other publicly available sources can be received on board and taken into account in voyage planning.

DISPATCH OF VESSELS

Where forecasts received indicate prolonged periods of adverse weather at the offshore locations to be supported on a particular voyage such that it is unlikely that any of the intended sites can be worked safely, the master(s) and charterer should agree that the dispatch of any vessels involved should be deferred until anticipated conditions improve. In the event that a vessel is dispatched in such circumstances the master may, at their sole discretion, elect to take an indirect route to reduce the risks to the ship, its personnel and cargo, or to proceed to a sheltered location to await an improvement in conditions the offshore locations. In this context "prolonged period" should be taken as period exceeding approximately one day (any 24 hour period or part thereof) where it is unlikely that any work could safely be undertaken at any of the relevant offshore locations in the forecast conditions.

POTENTIAL DROPPED OBJECTS

Unsecured objects being dislodged or falling from cargo items represent a risk to personnel, equipment and the environment throughout the supply chain. At all stages in the supply chain items should therefore be thoroughly inspected prior to transfer from one stage to the next. Potential dropped objects identified during these inspections should be removed and reported. Objects which constitute this risk include, but are not limited to loose tools used when servicing equipment included in or forming part of the cargo item; foreign objects in or on containers, including in fork lift pockets; and ice formed when water entrained in a cargo item freezes. When loading or discharging any deck cargo the personnel involved should therefore move to a safe haven well clear of the intended load path until it is safe to approach the item, or it is no longer above the vessel's deck. Where practical and safe to do so items on the deck should be inspected for potential dropped objects after loading and again before discharge at the offshore facility or onshore base. Any such objects identified during these inspections should be removed if safe to do so and an incident report submitted through the appropriate channels. If the object(s) cannot be safely removed the cargo item should be quarantined pending a full assessment of the risks which may be involved in discharging it, either at the offshore facility or quayside.

STOWAGE AND SECURING OF CARGOES IN CONTAINERS

Failure to correctly secure cargo items shipped in containers, either open or closed, can pose serious risk to personnel and equipment, including injuries being sustained by crew members when attempting to secure the loose items, and a change in the centre of gravity of the lift due to the movement of loose items within the container could result in its being significantly out of level. This may result in the loss of contents from the container. Handling of the load, particularly when landing, will also be made much more difficult. The proper packing and securing of cargo within any container is therefore a safety matter of the highest importance. Any person who has reason to believe that the correct procedures have not been followed or satisfactory arrangements installed should "stop the job" until remedial measures have been implemented.

REFRIGERATED CONTAINERS (DISCONNECTION AT THE OFFSHORE FACILITY)

From time to time refrigerated containers may be used to deliver provisions to offshore facilities. Such containers may have their own self-contained refrigeration unit, but more usually electrically powered units will require connection to receptacles on vessels which have been specifically installed for this purpose. Specific check lists may relate to the carriage of such items which should be completed by the relevant personnel. Where such container(s) are used it is important that they are not isolated for significant periods since the temperature may rise to such an extent that the contents thaw and have to be condemned. It is therefore recommended that when preparing to discharge this type of container at an offshore facility the power supply should be isolated, disconnected and removed only from those items to be delivered to that facility. The power supply may remain connected to refrigerated containers intended for other destinations. In some circumstances it may be necessary to isolate, disconnect and remove the power supply to those containers to be delivered to an offshore facility prior to entering its safety zone.

TUBULAR CARGO

General Guidance. General guidance relating to best practices when transporting tubular cargoes is provided at *Annex H (Carriage of Tubular Cargo).*

Round Tripped Tubular Cargo. It is recommended that when tubular cargoes remain on the vessel for successive voyages to an offshore facility the following practices are adopted to prevent incidents. Lifting arrangements should be checked to ensure that they are correctly installed prior to loading any other similar items "on top". Such checks should include: (1) correct leads of all parts of lifting arrangements; (2) presence and correct installation of securing arrangements such as bulldog grips, Velcro straps, tie-wraps, etc.; and (3) assessment of the adequacy and suitability of the above mentioned securing arrangements. Prior to lifting any bundles from the vessel deck at the offshore facilities a check should be made of BOTH ends of the lifting slings to ensure that they are correctly set up for the lift. Where appropriate a risk assessment of the discharge of such items should be undertaken and the outcomes included for discussion in the subsequent toolbox talk.

MAIN BLOCK OPERATIONS

Cargo items will normally be transferred between a vessel's deck and an offshore facility using the auxiliary hoist (otherwise known as the *whip line*) of the latter's crane(s). From time to time, however, where the weight of the item to be transferred exceeds the capacity of the auxiliary hoist the crane's main hoist must be used. Should this be necessary an intermediate pennant of sufficient safe working load should be installed on the hook(s) of the main block enabling personnel to connect or release the lifting rigging on the cargo item without having to approach or attempt to manoeuvre the block itself. Where practical, this intermediate pennant should be of sufficient length such that the height of the main block when the lifting rigging is connected or released is always approximately five metres (16 feet) above the cargo rails at the side of the main deck, or the highest adjacent item of cargo if this extends above the cargo rail. Any requirements to undertake operations of this nature should be advised to the vessel(s) involved in sufficient time for the appropriate task-specific risk assessments to have been made. Operations should not commence until the vessel has confirmed that these assessments have been completed and personnel briefed as to any particular precautions to be observed.

CHERRY-PICKING

"Cherry Picking" may be defined as being "selective discharge of cargo from within the stow". The term "cherry picking" generally infers cargo lifting arrangements not being directly accessible from deck level; breaking stow from an open location with no clear and secure access / escape route(s) to adjacent safe havens; and or any requirements for personnel to use unsecured ladders or to climb on top of other cargo or ship's structure and to enter any container to connect lifting arrangements is prohibited at all times. Masters who may be asked to undertake any of the above should "stop the job". To minimise the risk of "cherry picking" every effort should be made prior to commencement of loading to ascertain which, if any cargo items are of high priority. Vessels will be advised accordingly and cargo should be stowed in such a manner that any high-priority items can be discharged directly on arrival at their destination. Such cargoes are to be identified before cargo is loaded onto the vessel.

OTHER POTENTIALLY HAZARDOUS PRACTICES

The following practices may also be potentially hazardous and should be individually risk assessed. Moving other cargo on deck of vessel to gain access to a particular item. Lifting cargo containers to the deck of the facility, stripping the same and returning to the vessel, with the vessel being required to remain alongside the installation throughout. It should be appreciated that such practices may introduce increased risks due to additional lifting operations, involving increased risk to personnel; or vessels having to remain close adjacent to the facility for extended periods, involving increased risk of collision. Masters who are asked to undertake either of the above should challenge any such requests, drawing attention to the additional risks outlined above. Furthermore, before proceeding with any of the activities referred to above a thorough risk assessment should be undertaken and the outcomes included for discussion in any subsequent toolbox talks. Where frequent requests to support operations of this nature are received from a particular facility concerns relating to the risks involved should be raised with the facility manager and the charterer.

UNUSUAL CARGO ITEMS LOADED ONTO VESSEL DECKS

From time to time requirements may exist for unusual items to be loaded onto the deck of OSVs. Examples of such items include, but are not limited to modules or large fabricated items associated with offshore construction projects, very long items, including tubular cargo, flare booms, crane booms or similar, which, because of lifting geometry require the use of two stinger pennants on the crane hook, and any items which have not been pre-slung prior to shipment. Such items may have unusual dimensions, be unduly heavy or have high footprint loads, have unusual means of support and their transportation may have been the subject of a specific engineering assessment. In addition, connection and release of the lifting rigging may pose particular risks for personnel on the vessel. In this context any cargo items not carried in conventional shipping units such as containers, baskets, tanks or racks should be considered "unusual". The master of the vessel proposed for the carriage of any such items should be notified of the intention to load them on his vessel sufficiently in advance for the potential risks associated with their loading, carriage and discharge to be fully assessed.

TAG LINES

In general, it is recommended that the use of tag lines should be avoided. However, it is recognised that their use may be advantageous in handling some of the cargo items referred to above, and also that they are in general use in certain parts of the world. Guidance for their make-up and use is therefore included at *Annex I (Make Up and Use of Tag Lines)*.

Chapter 8

Bulk Cargo Operations

GENERAL REQUIREMENTS

Cargoes carried in bulk on OSVs include dry products in powder form together with various types of oil and water based muds, base oils, brine and numerous other chemicals transported in liquid form. Attention is drawn to chapter 1 *(Roles and Responsibilities)* which emphasises, that when planning to load any cargo, including those consisting of bulk powders or liquids onto an OSV, the various parties involved have several joint responsibilities, including ensuring that the proposed vessel is fully fit for the purpose intended; it is fully in compliance with all relevant legislation, rules and codes relating to the carriage of the relevant goods or products; appropriate procedures for the loading, carriage and discharge of the products are in place; the personnel involved have relevant experience and competencies; and, bulk cargo transfer is potentially hazardous and must be done in a controlled manner.

GENERAL PRECAUTIONS

In undertaking bulk cargo operations the following precautions should be observed. (1) The pressure ratings of all components of the transfer system should be verified to ensure that they are appropriate for the proposed operation. (2) Prior to commencement, agreement shall be reached between all relevant parties, including vessel, base, facility or roadside tanker regarding the pressure rating to avoid overpressure. (3) The protocols for control of the transfer operation are to be agreed by all parties involved. (4) Communications arrangements are to be agreed and to be tested prior to the commencement of the operation and at frequent intervals as it proceeds. (5) If communications are lost, *stop the job.* (6) The shipper and receiver should confirm quantities to be transferred and subsequently monitor at regular intervals. (7) The shipper and receiver should agree rates of delivery and densities of cargo being transferred. (8) Relevant personnel must be readily available and nearby throughout transfer operations. (9) At the facility the master or senior OOW must ensure they can see the bulk hose(s) at all times and not be distracted away from these. (10) Particular attention should be paid during hydrocarbon transfers that proper consideration is taken of potential hazards when carrying out concurrent cargo operations. Each party must give sufficient warning prior to changing over tanks and communicate when changes have occurred. Never close valves against a cargo pump. If at any point the vessel master, shipper, OIM or any other person has concerns relating to the safety of the transfer, the operation must be terminated immediately. Unregulated compressed air should not be used to clear any bulk hoses back to the vessel as this may damage the tanks. Compressed air should not be used to clear hoses used for the transfer of any hydrocarbon based products since an increased risk of explosion will result. Never transfer any other liquids using potable water hoses. Before use, flush potable water lines through to clear any residues. Hoses must remain afloat at all times through the use of sufficient floating devices. The use of self-sealing weak link couplings in the mid-section of the hose string is strongly recommended. Always avoid the use of heavy sections of reducers or connections at hose ends. The hose from the facility should not be connected to the vessel until both have agreed that all preparations have been completed and that the transfer can commence immediately after connection has been satisfactorily completed.

BULK OPERATIONS IN PORT AND AT THE FACILITY

Flow charts illustrating the processes involved in handling of bulk cargoes both in port and at the offshore facility are included at *Annex T (Flowcharts for Handling Bulk Cargoes in Port and Offshore)*. Particular responsibilities associated with such operations are described below. A check list which should be completed prior to commencing any transfers of bulk cargoes is included at *Annex J (Wet Bulk Cargo Transfers)* and *Annex K (Dry Bulk Cargo)*.

Vessel responsibilities at the facility. Before offloading bulk cargo, it is necessary for the vessel and facility to confirm a number of key parameters, including: communications protocols with the receiving facility; whose "STOP" it is; the quantity of bulk to be offloaded; the necessary hoses and connections, colour codes and dimensions; whether the rigged hose lengths are adequate; the procedures for venting and blowing through the hoses; whether the facility is ready to receive the cargo; whether all valves and vents are open and the correct tanks are lined up; and whether the relevant emergency shut down procedures are in place and the crew are familiar with same. It is also necessary to ensure that all pollution prevention equipment is in place as per the SMPEP and all manifold valves must be in good condition. Moreover, the PIC cannot be distracted from the operation and any facility under-deck lighting must be adequate. Dry bulk vent line positions must be identified and, where necessary, prepared. The master should submit to the designated contact person all receipts where applicable, including meter-slips, for the cargoes transferred in addition to any other relevant documentation and information.

Facility responsibilities. The facility must ensure that communications protocols have been agreed and, in particular, whose "STOP" it is. Hoses, manifolds and valves must be visually inspected, maintained and replaced as required and or in accordance with the planned maintenance system. Slings and lifting points must be visually checked and replaced as required. Hoses should only be lifted by a certified wire strop on a certified hook eye fitting. Under-deck lighting must adequately illuminate the transfer hose and vessel. Appropriate flotation systems must be intact and in place.

PREPARATIONS RELATING TO THE TRANSFER OF DRY BULK MATERIALS

The following recommendations are included to supplement those in the flow-charts included at *Annex T (Flowcharts for Handling Bulk Cargoes in Port and Offshore)*. It is recommended that procedures should be adopted as follows. Prior to confirming that a vessel is ready to transfer any dry bulk cargoes it should be verified that all on board preparations have been completed. This includes a requirement to ensure that, where relevant, all elements of the system have been vented to atmospheric pressure. When transferring dry bulk cargoes to or from vessels, the personnel responsible for delivering the product should confirm that those responsible for receiving it have completed all relevant preparations. Assumptions that preparations have been completed can be dangerous and must be avoided at all costs. Relevant check lists are to be completed as required by the parties involved. When transferring dry bulk cargoes to or from vessels, care should be taken when deciding the sequence and manner in which the various valves are opened to avoid the risk of inadvertently over-pressurising any elements of the system. It will be appreciated that the handling of dry bulk materials involves systems containing large volumes of pressurised air. The stored energy in such systems is therefore considerable and the potential for serious personal injury in the event of failure is high. All personnel involved in such operations must therefore comply with all relevant procedures and to ensure that all checks have been satisfactorily completed prior to confirming their readiness to deliver or receive the product.

HOSE USAGE

General guidelines regarding the usage and care of offshore bulk hoses are included in *Annex N (Bulk Hose Best Practice)* and *Annex O (Bulk Hose Handling and Securing)*.

HOSE MARKING CONNECTIONS

Further information relating to hose marking, usage and connections are included in *Annex P(Hose Markings and Connections)*.

BULK HOSE HANDLING PROCEDURES AT THE FACILITY

It is recommended that the following procedures be adopted during receipt and handling of bulk hoses at the offshore facility. The vessel should take up position and confirm readiness to receive the hose. Except where other arrangements are in use, the crane operator on the facility lowers the hose to the vessel, holding the hose against the ship's side and at a height that allows the crew to catch and secure it to the vessel's side rail, keeping the hose end clear of the crews' heads. Where other arrangements exist the appropriate procedures should be followed. Once secure, the hose end is lowered inboard of the rail and the crane hook disconnected. When the hook is clear, the crew install the hose on the appropriate connection on the ship's manifold. Uncoupling is the reverse of the above procedure. After releasing any self-sealing connection it should be visually inspected by the deck crew to ensure that it is fully closed and is not passing any liquids. Vessel crews should be reminded that hose couplings should, whenever possible, avoid contact with the ship's structure. The integrity of the couplings should be monitored by visual inspection of the painted line on the couplings, where applied. In marginal weather greater care than normal is needed by the vessel to avoid over running the hose especially if deck cargo is also being worked. Consideration should therefore be given to working bulk only in such circumstances.

HOSE SECURING ARRANGEMENTS

Section 10.7 of the Guidelines describes the general principles of handling bulk hoses at offshore facilities, however, these often require personnel securing the hose(s) to work underneath or in very close proximity to the suspended hose for some time whilst completing the arrangements used to secure the hose. Any arrangements which reduce the time personnel are required to work in close proximity to the suspended hose, or avoid this altogether should be investigated. *Annex N (Bulk Hose Best Practice)* provides a description of arrangements which are relatively simple, requiring only minor modifications on any vessel and none on the facility other than the rigging of a soft strop at a suitable distance from the end of the hose. Whilst, when it is being passed to the vessel personnel still have to work in close proximity to the suspended hose when securing it, the time required is much reduced. On recovery, personnel are required to disconnect from the manifold and connect the crane hook to the recovery pennant, but thereafter further intervention is not normally required. Other arrangements having similar objectives have also been developed. Further proprietary arrangements have been developed, for example those illustrated at *Annex O (Bulk Hose Handling and Securing)*. These also involve minimal modifications to the vessel and have been used successfully in some operational areas. Whilst undoubtedly minimising risks associated with hose handling even further such arrangements are likely to be more complex, requiring fairly extensive modifications on both vessel and facility. The flexibility to utilise vessels not equipped with the particular features required will therefore be reduced.

BULK TRANSFERS OF COMMON LIQUIDS

There are a number of procedures that must be followed when preparing for, or carrying out bulk transfers of common liquids, which are briefly discussed here. *Cargo Fuel (Marine Gas Oil)*. Establish a sampling and receipting procedure when transferring fuel. Sampling taken in accordance with *MARPOL Annex VI* will normally suffice for these operations. However, in some circumstances more rigorous sampling procedures may be required. Any such requirements should be included in the master's sailing instructions and should always be complied with. *Potable Water*. Specific national or charterer's requirements may apply to the carriage, storage and transfer of potable water. The charterer, owner and master should ensure that any such requirements are understood and fully complied with.

BULK TRANSFERS OF SPECIAL PRODUCTS

Special care must be taken to follow the correct procedures when transferring special products which include, but are not limited to, methanol and zinc bromide. Appropriate risk management procedures should be in place when transferring special products. Reference should be made to chapter 2 *(Operational Risk Management)* with particular attention being given to the provision of personal protective equipment required for personnel involved. When transferring these products the following should be observed. *Shipper*. Provide full details of the product(s) being shipped, including details of all precautions to be taken when handling said products. Staff to be on site throughout to advise on pumping, handling, earthing and the discharging of tanks. Provision of appropriate fire fighting equipment, where relevant. *Operating company and base operator*. Nomination of berth after liaising with the harbour authority, fire brigade and harbour police or security. Ensure sufficient cooling or drenching water is available. Cordon-off area, with signs posted to indicate a hazardous area. *Master*. Should complete a ship to shore safety check with the shipper. Must authorise loading. If required, ensure a permit to work is in place before any loading operations are performed. Ensure the vessel's restricted zone is clear, fire hoses are rigged and SMPEP equipment is ready for action before commencing loading.

Characteristics of some Special Liquid Products. Whilst the shipper should provide full details of any products being shipped, some of the main characteristics of the more common chemicals which may be shipped in bulk liquid form are included here. *Methanol*. Particular characteristics of this product are it burns with no visible flame in daylight conditions; is readily or completely miscible with water; is a class 3 substance with a noticeable odour; is highly flammable, with a flash point below 23°C (73.4°F); can evaporate quickly; has heavier than air vapour that may be invisible, and disperses over the ground; can form an explosive mixture with air, particularly in empty unclean offshore containers; experiences pressure increase on heating, with the risk of bursting followed by explosion; is very toxic, and possibly fatal, if swallowed or absorbed through the skin. Symptoms may not appear for several hours; and, can cause significant irritation of the eyes. The following specific precautions should be observed when transferring methanol. Ensure that the integrity of the system is intact, including all relevant certification which should be valid and in-date. During bulk methanol transfer, smoking and the use of ignition sources are strictly prohibited. During electrical storms (lightning) operations should be terminated. Free deck space around bulk loading / discharge stations so that coverage of foam monitors is not obstructed. No other operations to be undertaken when handling this product.

Zinc Bromide. Zinc Bromide is a highly corrosive and environmentally contaminating product. Due to its corrosive nature, protection against injury from exposure to it is essential. Information provided by the shipper should be used when undertaking risk assessments involving the carriage of this product to determine the appropriate level of personal protective equipment which should be used.

ATTENDANCE OF FACILITY PERSONNEL DURING BULK TRANSFER OPERATIONS

Whilst vessels are connected to offshore facilities by hose(s) for the purpose of delivering bulk commodities to facilities it is important that, in the event of a change in the operating circumstances developing, personnel on the facility remain available at all times to disconnect the hose(s) at short notice. Failure to disconnect the hose(s) in a timely manner should circumstances change during bulk transfer operations could well result in significant risk of injury to personnel and or damage to assets or the environment. The crane operator and deck crew on the facility shall therefore remain readily available, contactable and nearby throughout transfer operations. In the event that any such personnel are required to leave the vicinity of operations for any reason the vessel should be immediately advised. The vessel bridge team in conjunction with the facility manager should assess current and anticipated operational risks. It is the master's decision as to whether the vessel remains connected to the facility pending restoration of the required level of support.

BACK-LOADED LIQUID BULK CARGOES

Refer to *Annex Q (Carriage of Oil Contaminated Cargo)* for further details.

TRANSFER OF NOXIOUS LIQUIDS DURING THE HOURS OF DARKNESS

It is recognised that it will be necessary to transfer hydrocarbon or other noxious liquids during the hours of darkness, particularly in higher latitudes in the winter months. For clarity, the Guidelines do not advocate that such operations should be curtailed or restricted, but instead seek to identify the additional risks involved in such transfers and to make appropriate recommendations to manage such risks. It is recognised, for example, that leaks are most likely to occur in the early phases of any transfer operation as connections become pressurised. Once all aspects of the transfer operation have been stabilised leaks are less likely to occur. It is therefore recommended that, wherever practical, the following practices may be adopted in relation to the bulk transfer of hydrocarbons (or other recognised marine pollutants) during the hours of darkness: adequate artificial illumination of the operational areas on the facility, the vessel and the water between them should be provided; additional high-visibility and or reflective panels on the hoses (or their buoyancy elements) are recommended; all preparations for the transfer to completed in daylight, where practical; careful check to be made for leaks, etc. on vessel, facility and connecting hose as transfer commences; transfer may continue into the hours of darkness, provided that the entire area and associated equipment is adequately illuminated to an acceptable standard; in the event that the transfer continues a careful watch of the connections and hose should be maintained throughout; it is recommended that hydrocarbons or other noxious products should not be transferred simultaneously in these circumstances; on completion of the transfer extra care should be taken when breaking the connection and returning the hose to ensure that the risk of spillage on completion of the operation is also minimised. General precautions to be observed regarding safety of personnel working on deck during the hours of darkness should continue to be implemented.

TANK CLEANING

Preparations

Risk assessment. The tank cleaning foreman must demonstrate to the master that they understand the principles and, if necessary, has undertaken a risk assessment relevant to the intended task. The outcomes of the risk assessment should have been addressed in the subsequent toolbox talk prior to commencing the task. *Protective Equipment.* Personnel working in the tank must wear the appropriate personal protective equipment as identified in the risk assessment, COSHH, or equivalent assessment and MSDS. *Atmosphere Testing and Tank Entry.* All tanks should be considered as "dangerous spaces" which, if appropriate precautions are not taken, would represent a serious risk to personnel entering them. The tank cleaning foreman must demonstrate to the master that the atmosphere in the tank has been tested to prove that it does not represent a threat to any personnel who may be required to enter the space. They must also be able to demonstrate that any equipment utilised for this purpose has been used in accordance with the original equipment manufacturer's instructions. The results of the atmosphere testing should be recorded on the permit or other agreed document. *Communications.* Communication systems between all personnel within the tank and at the point of access must be agreed, tested prior to commencement of cleaning activities, and checked at frequent intervals until all persons have exited the tank on completion of operations. A stand by person at each tank entrance will almost always be required. This person should be competent and trained to take the necessary action in the event of an emergency. Effective means of ship to ship and ship to shore communication must be established and maintained throughout the tank cleaning operation. *Emergency Response and Escape.* The tank cleaning foreman must demonstrate to the master that the emergency response and escape arrangements identified in the risk assessment are in place and available if required. *Check List.* A typical example of a check list which should be completed prior to the commencement of tank cleaning operations is included in *Annex L (Tank Cleaning Check List)*.

Operations

Control. Although the tank cleaning operation is conducted by a contractor under control of the contractor's supervisor the safety of the operation remains the responsibility of the master. The operation should be continuously monitored by a designated responsible vessel person who should stop any operation that they consider unsafe. *Atmosphere Testing.* Regular tank atmosphere testing by competent personnel from both the vessel and the tank cleaning contractor must be undertaken both prior to the commencement of cleaning activities and checked at frequent intervals until all persons have exited the tank on completion of operations. Equipment utilised to conduct these tests of the tank atmosphere must be used in accordance with the original equipment manufacturer's instructions. *Simultaneous Operations.* Where simultaneous tank cleaning and other operations, i.e. cargo operations, are undertaken, then suitable safety precautions must be in place. Interfaces between the vessel's officers, tank cleaning supervisors and quay supervisors must be kept open and active throughout the tank cleaning operation. *Shift Hand-Overs.* Hand over between shifts of the vessel's and tank-cleaning personnel must be carefully controlled to ensure continuity. Consideration must be given to holding a further toolbox talk.

Completion of Tank Cleaning

On completion of the tank cleaning operation the master must carry out an inspection together with the tank cleaning contractor supervisor to ensure that the tanks have been properly cleaned and lines and pumps are thoroughly flushed. If these parties disagree an independent surveyor will carry out an inspection. The various commonly accepted tank cleaning standards are shown at *Annex M (Tank Cleaning Standards)*. The tank inspection should confirm that the tanks have been cleaned to the appropriate standard.

Chapter 9

Anchor Handling and MOU Moving

MOU moving operations are potentially hazardous and all personnel should appreciate their joint responsibilities as defined below in the section *Jointly Agreed Procedures and Responsibilities*. As such, the guidance in this chapter applies equally to all types of MOUs.

DEFINITIONS

MOU. MOU is defined as for example, but not limited to MODUs (all types), FPSOs, barges, accommodation units (all types), and self-propelled, self-elevating service units.

Person in Charge (PIC). The PIC is to be identified. The person supporting this function will be described in the unit owner's operating procedures, and may change dependent on the operations being undertaken. Any changes in the persons in charge of the operation should be recorded in the unit's log book, being counter-signed by both the individuals involved. To avoid any potential conflicts of interest tow masters and operator's marine representatives should, wherever possible, be sourced from different independent contractors. The responsibilities and authorities of the principal personnel involved are to be clearly defined.

Vessel. Vessel(s) in the context of this chapter means anchor handlers or tugs used to assist the MOU to move location and to work on, deploy or recover MOU moorings.

RESPONSIBILITIES

Jointly agreed procedures and responsibilities. The parties involved will agree who will be responsible for the preparation of the Work Specification. In most instances this will be the unit owner, who may delegate this task to an independent contractor acting on their behalf. All parties are jointly responsible for ensuring adequate planning (including contingencies). They should also agree on the risk management procedures to be observed, and are jointly responsible for ensuring that this is complied with throughout the entire operation. A management of change process should be included in the risk management procedures. Any deviation from the Work Specification shall only be permitted in accordance with this agreed management of change. Reference should be made to chapter 2 *(Operational Risk Management)* relating to the risk management process. Each party involved should determine how its interests will be represented during the operation, and ensure that others are advised of the relevant arrangements. A Work Specification that covers the entire operation shall be prepared. The Work Specification should be written in English unless otherwise agreed. Identify who will have the responsibility and authority to specify necessary equipment in accordance with the Work Specification. Ensure that satisfactory anchoring and mooring analyses have been prepared in compliance with national and industry requirements where relevant. As early as practicable organise pre work scope meeting, to include risk assessment if required. Ensure that the work scope has been reviewed and is understood by all personnel that participate in the operation. Agree party who is will charter vessels and mobilise according to the work scope. Arrange inspection of selected vessels to verify suitability in accordance with marine assurance protocols agreed between the parties involved. Inform vessel(s) and MOU about the status of the operation at all times. Any proposed personnel changes during MOU moving Operations to be arranged so that relief personnel have sufficient time to be fully briefed on status of work scope by those they are relieving. Communicate any changes of the work scope to all the parties involved.

Responsibilities of the operating company. This section relates to the organisation or company contracting the MOU. The operating company should obtain the location information necessary to moor, un-moor, emplace or extract a MOU at or from the relevant location(s). Obtain an overview of infrastructure on the seabed, sea bottom conditions and any obstructions. Provide charts and location layout drawings showing intended MOU position(s). Provide the necessary charts and drawings in both paper copies and in electronic format. Specify minimum horizontal and vertical clearances to be maintained from such as infrastructure, adjacent moorings and pipelines on the seabed. Ensure that third parties having interests or assets in the vicinity of operations are advised of intended activities, and invited to participate in operational meetings or risk assessments should they so wish. In addition to the above the operating company shall make all the vessels available for a common briefing, ideally in port, prior to mobilisation. This briefing should be attended by the master(s), the vessel's officers and the deck crew of the vessel along with MOU owner's representatives.

Responsibilities of the MOU owner. Notify authorities of MOU departure and arrival in accordance with local requirements. Ensure MOU is adequately manned by competent personnel taking into account hours of rest requirements and the scope of work. Provide extra personnel as required to cover 24 / 7 operation. Ensure arrangements for provision of additional / back-up mooring equipment, if required, are in place.

Responsibilities of the PIC of the MOU. The PIC has overall responsibility and the authority for the HSSE management of the facility and personnel at all times as per statutory requirements and MOU owners' policy. However, in relation to the movement of the MOU operational responsibility may be delegated to a suitably qualified person such as the tow master who should also consult with the vessel master in the process. Such responsibilities include, but are not limited to, the following:

- Decisions as to when it is safe and practicable to commence operations within the limitations of the MOU operating manual, having consulted with the operators representative and the vessel(s) master(s).
- Ensuring that a meeting is held with all relevant personnel on board prior to operation and minutes accordingly, with an appropriate entry in the log book to that effect.
- Masters of Anchor handling vessel's in attendance to support the move should be advised of the outcomes of this meeting and invited to comment on these.
- Ensuring procedures in place to monitor each vessel's operation, and to monitor ongoing status of the operation.
- Ensure function and monitor for effective communication between all involved parties.

- Acts as the sole point of contact through which all operation notifications and exterior communications will pass and ensures that all relevant authorities are kept informed of the operation, as required.
- Liaises and communicates with the operating company representative on all matters concerned with the operation and any deviation from the agreed work scope.

Responsibilities of the vessel owner. Responsibilities of the vessel owner include, but are not limited to, ensuring that vessels are in good operational order and in compliance with relevant legislation and charter party requirements. Ensuring that vessels are manned by competent personnel taking into account hours of rest requirements and scope of work including possibility of 24 / 7 working. Ensuring that any proposed personnel changes during MOU Moving Operations are arranged to allow sufficient time for a briefing on work scope and experience transfer to be completed. Ensuring that the vessel is able to calculate and monitor stability information for all stages of the intended operation. Ensuring that ship specific anchor handling manuals or procedures are included in each vessels' safety management system and that such documents are available on board. Ensure that a clearly defined clear-deck policy, when equipment under tension is present, is understood and implemented on each vessel. Ensuring that details of the vessel(s) provided to brokers and charterers are correct and current.

Responsibilities of the vessel master. The prime responsibility of the master of any vessel is to safeguard the safety of the vessel's crew and equipment on board and the marine environment at all times. The master is empowered to stop operations that may put personnel, vessel or environment at risk. In fact, *all* personnel have the right to call a stop at any time. In multi vessel operations masters should avoid allowing perceived "peer" pressure to influence the overall decision making process. Other responsibilities include, but are not limited to:

- Ensuring that the manning provision on board is sufficient based on working hour provisions, the operation work scope and that the crew is rested.
- Ensuring that all anchor handling equipment on board or supplied, is in good condition and certificated as required and meets the requirements of the Work Specification, including, where relevant, the vessel's own equipment.
- Defects or non conformities to the anchor and or mooring equipment found during the operation are to be reported to the charterer.
- Ensure compliance with the vessel owner's and charterer's HSSE policies relating to the reporting to the PIC on the MOU of any accident, incident or vessel deficiency / limitation occurring during the operation.
- Ensuring that a Vessel Risk Analysis has been performed and recorded, in accordance with the specific work scope, and ensure that the agreed work scope is communicated to all crew members involved in the operation.
- Ensuring that the stability of the ship is calculated and recorded, for each step in the work scope, including worst case expected dynamic loads.
- Ensuring that sufficient consumables are on board for the intended operation.
- Ensuring that the personnel who will be involved in the operation have been adequately briefed as to the nature of their duties and responsibilities.

PRELIMINARY PREPARATIONS FOR THE MOVE

Work specifications. A detailed written Work Specification should provide all necessary information for proposed operations and describe them in detail. It should provide common understanding of the operation and should outline framework conditions, using images, animations, organograms and diagrams where possible. It is intended for use during the planning, execution, verification and demobilisation of the operation. The suggested contents of the Work Specification are included at *Annex R (Guidelines for the Content of MOU Move and Anchor Handling Work Specifications).*

Contents of the Work Specification. The Work Specification should include, but not be limited to the following information:

- Identification of key roles, responsibilities and agencies involved.
- Defined health, safety and environmental expectations. Reference should be made to chapter 2 *(Operational Risk Management)* relating to risk management process.
- Statements of HSSE reporting requirements / expectations by all parties.
- Identification of set trigger and hold points which determine operation start, stop, and hold or Risk Assessment.

Examples of potential trigger points include:

- Significant sea height reaches *xx* metres[1].
- Wind velocity reaches *xx* knots[1].
- Current velocity reaches *xx* knots or adversely impacts on operations[1].
- Reduced Visibility.
- Unexpected loads experienced either by any Vessel or the MOU.
- Mooring equipment problems.
- Any technical problems aboard any Vessel or MOU.
- Any technical faults with the survey equipment.
- If at any stage there is any doubt about being able to maintain the clearance between the deployed chain or wire catenary and any sub sea asset.

Examples of potential hold points include:

- Prior to recovering Secondary Moorings.
- Prior to recovering Primary Moorings and Going on Tow.
- Prior to entering safety zone at destination location.
- Prior to manoeuvring / mooring operations at destination location.
- Prior to running secondary moorings.
- Prior to moving along side or over another structure.
- Prior to commencing simultaneous operations.
- Show clearly the order of work and method to be utilised.
- List of equipment required during intended operations.
- Maximum calculated loads and dynamic tensions likely to be experienced by any vessel during the course of operations.

Examples of detailed drawings include:

- Anchor pattern.
- Each stage of operation when moving off or on to the location.
- Make up of mooring lines, including any extra equipment in use.
- Details of length and connection types are to be included.
- Passage plan to be defined and agreed by relevant parties.
- Significant sea height, wind and current velocities to be determined by operational considerations or risk assessment.

Guidance on Cross Track Distances. It is advisable to keep forces being exerted by the weight and tension of the mooring system along the fore and aft line of the vessel(s) by keeping the vessel(s) on, or close to, the intended mooring line track and bearing. Due consideration must be given to any and all effects of any deviation and the prevailing environmental conditions. During the planning of the Work Specification loads will be identified to show the expected forces on the mooring systems and the respective distance from the MOU and cross track limits set. For example when Vessel range from MOU is less than 160 metres (524 feet) no action is required as loads are low and the vessel may be required to deviate off line to work crane, rack anchor or position correctly for environmental forces. Alternatively, *Table 9-1* sets out the actions to be taken when the vessel range exceeds 160 metres (524 feet).

Table 9-1 Cross Track Distance Limits and Actions

ZONE	LIMITS	TITLE
Green	> 150m each side of intended track	No action required
Amber	Green zone + 150m on each side of the intended track	1. Vessel instructed to regain line 2. Assistance from MOU provided if required 3. Review environmental forces being experienced
Red	Outside green and amber zones	1. Mooring operation suspended until vessel regains amber zone and movement toward intended track confirmed 2. PIC notified

Pre-Operation Meetings (Onshore)

Participants. The pre-operational meeting onshore should be held well in advance of the commencement of operations. Written Work Specifications for the MOU move should be available and agreed with all relevant parties. Appropriate competencies should be made available for this meeting and are likely to include the persons in the following roles:

- The OIM or MOU designate (either in person or participating remotely).
- The tow master (preferably nominated for attendance on board MOU during the operation).
- Representatives from the MOU owner's operations department. (Note that in some jurisdictions the MOU safety delegate may also be required to attend).
- The onshore and or offshore supervisor for the operating company.
- Onshore logistics representative from the operating company.
- Marine representative / superintendent from the operating company.
- Representatives from the navigation, positioning, survey company or contractor.
- If required, the operating company's navigation, positioning or survey representative.
- Representatives from the owner / operator of any third parties having interests in the vicinity of the proposed operations.
- Representatives of the owner(s) or masters of the vessel(s) involved, if on charter at time of meeting or where options exist for employment at some future date. This may relate to some particularly complex moves for which particular vessels are required.

Meeting Agenda. As a minimum, the agenda of the pre-operational meeting should include:

- Confirmation of responsibilities and authorities.
- Review of the Work Specification.
- Confirmation that onshore pre-move HIRA has been completed, with outcomes available for discussion during the meeting. If this is not the case, the HIRA should be included on the agenda.
- Confirmation of who is to provide weather forecast data, wave, tidal stream and current data.
- Confirmation of anticipated loads which may be experienced by vessels are included in the Work Specification.
- Confirmation of positioning equipment and positioning personnel.
- Determination of logistical needs for completing the Work Specification.
- Confirmation that all equipment to be utilised is available, has been confirmed fit for purpose by inspection, and certified.
- Review of risk assessments for operation and transfer of experience.
- Weather and environmental limitations and definition of operational criteria, review and

set trigger and or hold points as required.
- Confirmation of navigational data and references.
- ROV inspection requirements, where relevant.
- Review of details of anchor pattern and mooring equipment inclusive of maximum calculated loads and dynamic tensions and any pre-tensioning requirements.
- Pre-lay of anchors, if applicable.
- Review of tandem vessel operations, if applicable.
- Confirmation of vessel requirements, including manning, quantity and technical specifications.
- Manning requirements on the MOU.
- Anticipated duration of each phase of the operation.
- Contingency plans and equipment.
- Equipment lists for the individual vessels.
- Sea bed conditions.
- Communication lines, and contact details.
- Defining of responsibilities and authorities for reporting during and debriefing after the completion of MOU operation. This should include lessons learned.
- Review of MOU Move and MOU Operational Equipment General Guidance.

Vessel Equipment

Towing and Anchor Handling Equipment Register. Apart from the maintaining its own statutory, and company logs, it is recommended that a vessel specific towing and anchor handling equipment register be maintained to record the status of certification, maintenance, hours and type of use. These records are to be maintained for all current equipment. This register should include, but not be limited to:

- Tow Wire, and spares if carried.
- Work Wire(s).
- Chasing Pennant(s).
- Pig Tail(s).
- Chafe chain(s) or similar arrangements.
- Stretchers, and Surge Chains.
- Shackles and Joining Links (of all types).
- Swivels.
- Running sheaves.
- J Hooks, and Chasers.
- Grapnels.

Inspection of Equipment. It is recommended that all anchor handling equipment, particularly wires used in the course of operations are inspected after each use. Records of inspection should be included in the register. It should be noted that soft eye pennants wear more quickly than hard eye pennants and require more frequent inspection.

Preparations for Operations. During any operation, as a minimum, the following provisions should be in place:

- All equipment to be maintained and operated in accordance with the original equipment manufacturer's instructions.
- Oxyacetylene or similar cutting gear, with adequate consumables to be available for immediate use.
- Operational mechanical stoppers or similar means to safely and effectively secure wire pennants, recognising likely loads on the wire and the load-bearing capacity of termination used.
- Pelican hooks or similar arrangements which involve personnel working in close proximity to wires under tension introduce unacceptable risks and should not be used.

- Alloy ferrule terminations are not considered suitable and should not be used.

Care must be taken when opening wire coils, in particular pendant wires. Turntables should be used (if available) as coils springing open following release of securing bands may cause injury. Vessel crews should ensure that stowage of anchors and equipment should be secured in the course of operations and alert to the risk of movement when securing arrangements are released. Certain types of anchor are inherently unstable and may fall over on moving decks. A risk assessment as described in chapter 3 *(Certification, Training, Competency and Manning)* should be completed prior to such anchors being loaded onto the vessel, deployed or recovered.

Vessel MOU Anchor / Mooring Equipment

Good guidance relating to the handling and use of equipment is also available from the various anchor and or equipment manufacturers or suppliers including permanent chaser pendant / pennant systems, anchor chain mooring connections providers, pendant buoy system providers, working wire chaser terminations on vessels, and back-up mooring systems.

THE OFFSHORE OPERATION

Pre-Operation Meetings. A meeting should be held on board the unit prior to the commencement of moving operations. The primary function of this meeting is to review the Work Specification for the MOU move and related HIRA reports to ensure that all relevant parties are in agreement with the procedures proposed.

Participants. This meeting should be repeated if personnel and or the Work Specification should alter. Except where otherwise agreed it should be presumed that operations will continue on a 24 / 7 basis. Recommended participants for this meeting should therefore include, but are not necessarily limited to the OIM and their delegates, tow master(s) (will normally act as Chairperson), Offshore Supervisor(s) for the operating company, marine representative(s) for operating company, MOU marine personnel, party chief from the navigation, positioning, survey contractor, representative(s) from the owner and or operator of any third party interests if attending on board the unit, and any additional specialist personnel as required. Attendance at the meeting should, however, take into account requirements for adequate rest periods. The Chair should ensure that the meeting is minuted for reference and records. Following the meeting, masters of Anchor handling vessel's in attendance to support the move should be briefed on its outcomes by the senior tow master and invited to comment on these.

Meeting Agenda. As a minimum the agenda of the pre-operational meeting should include:

- Confirmation of responsibilities and authorities.
- Confirmation of work scope.
- Confirm receipt of weather forecast data, wave, tidal stream and current data.
- Confirm with the vessel(s) master(s), details of maximum calculated loads for the operation.
- Confirm the readiness of positioning equipment and personnel.
- Review of the schedule, logistical needs and notifications of interested parties or external agencies.
- Confirm state of readiness of the vessel(s), MOU and equipment.
- Confirm that on board risk assessments for intended operations has been completed and outcomes passed to Supervisors for inclusion in subsequent toolbox talks.
- Trigger and hold points set or confirmed.
- Confirm ROV inspection requirements.
- Confirm tandem vessel operations, if required.
- Confirm vessel roles and responsibilities, and operation sequencing to be reaffirmed.
- Communication lines to be confirmed.
- Responsibility and required attendees to be defined for report or debrief of completion of MOU move operation, should include lessons learned.

Reporting and notification. Where required, the PIC, in cooperation with the vessel(s) and marine representative or tow master, should report to the appropriate national and local authorities, adjacent facilities and to local operations as required as detailed in Work Specification.

TOWING OPERATIONS

Passage planning and navigation. The master of the towing vessel is responsible for the preparation of a detailed passage plan and the subsequent safe navigation of the tow. Where appropriate, navigation warnings should be broadcast by the lead tug at regular intervals. The passage plan must be forwarded to the tow master on the unit and the master(s) on any other towing vessels for review and or comment prior to the commencement of operations. Where more than one vessel is utilised to tow the unit a lead tug will be nominated by the tow master. The master of this vessel will assume the responsibilities described above and shall also ensure that the other vessel(s) involved comply with the plans. This does not, however, relieve the master of any vessel from the responsibility of safeguarding the safety of personnel and equipment on board own ship.

Operational planning. The passage plan must be carefully developed with regard to water depth, other offshore and sub sea facilities, and emergency locations or refuges which may be utilised if required. Close attention should be paid to the length and catenaries of the tow wire and its relation to environmental conditions, water depth and vertical clearance over any subsea assets in the vicinity of any location or whilst on passage. Route must keep safe distance from any other facilities. Pass on the side that best assures tow will drift away from the facility in case of power loss or loss of tow. The passage plan shall not use facilities as way points. Regular weather forecasts are essential for operations of this nature. Normally two forecasts, twice per day from independent providers will be provided. Communication lines as agreed during the pre-operational meetings should be observed. Requirements for support vessels should be assessed. Typically, support vessels' tasks include, but are not limited to monitoring and plotting ship traffic along the towing route; use of all available means to warn vessels whose course is approaching the tow too closely or impeding its progress in contravention of the COLREGS; checking agreed destination location is clear and unobstructed before MOU approach and arrival; and functioning as reserve or contingency towing vessel, especially in adverse or deteriorating weather. Retrieval arrangements for the recovery of the main towing gear in the event of its failure should be fully operational. The MOU's emergency towing system should be rigged and ready for immediate use. A safe procedure for passing this system to a towing vessel in all weather conditions should be agreed beforehand. Where this involves the mobilisation of any additional equipment this should be on the unit and checked to be fully operational. Where an additional vessel is available as reserve towing vessel on passage, this should be rigged for towing.

The towing operation. The requirements below relate when direction of the tow has been transferred from the tow master to the lead towing vessel. The towing operation will conclude when direction of the tow has been returned to the tow master from the master of the lead tow vessel. The transfer of direction should be agreed via radio and the times logged by both MOU and the lead tow vessel. Manoeuvring operations on and off locations are always directed by the tow master.

Exchange of information during the towing operation. The MOU is required to monitor the following and report any changes to the lead tow vessel compliance with COLREGS on the unit, including lights and shapes on the tow; towing connections; weather conditions and forecasts; integrity of the unit, if relevant; MOU propulsion assistance. The vessel is required to monitor the following and report any changes to the MOU: compliance with COLREGS; tow line, particularly prevention of any chafing or friction. Either use towing sleeve, or regularly adjust wire length; on passage towing speed and heading, alterations to be made in a controlled manner; deviations to passage plan; adjustments to power output; when adjusting tow line length reduce engine power if required to avoid damage to tow line; if towing a MOU on anchor chains, the MOU may pay out chain to provide the optimum towing catenaries; total tow length and catenaries profile in relation to water depth should be calculated and any changes communicated between MOU and tow vessel; whilst towing, clear deck policies to be observed. Should any urgent or emergency work be required on the after deck whilst towing this should be fully risk assessed; if adverse weather expected the master to consider whether use of gog wire to control towing wire would be prudent or advantageous; when towing in adverse weather, dynamic forces are significant. Exercise great caution, particularly when there is a following sea; and, towing logs are to be maintained by the vessel.

Record of towing operations. Vessels engaged any towing operations should maintain complete records of such activities. Particulars of the information to be recorded are included in section 12.2.3.3 of the Guidelines.

ANCHOR HANDLING OPERATIONS

The vessel master is responsible for navigation of the vessel during anchor handling operations under the direction of the PIC on the MOU. Requirements for tandem anchor handling operations are to be described in the move procedures and are to be fully risk assessed. Where such operations are to be used the lead and supporting vessels will be nominated prior to commencement. Any change in these roles should be the subject of management of change review.

Anchor handling operation planning. The Work Specification must be carefully developed with regard to water depth, other offshore and sub sea structure(s) and MOU skidding plans to maintain agreed clearances. During the operation close attention should be paid to Cross Track distances. Close attention should be paid to the length and catenaries of the mooring and its relation to the water depth and any sub surface structure. Regular weather forecasts, normally from two independent sources, should be arranged. Communication lines as agreed during the pre-operational meetings should be observed. Assess what support vessels are required. Support vessels' tasks include, but are not limited to:

- Monitoring and plotting ship traffic within proximity of mooring site.
- Use all available means to warn vessels whose course is approaching the scene of operations too closely resulting in risk to themselves or other vessels.

The offshore anchor handling operation. Reference should be made to the following documents:

- Work Specifications for the relevant operations.
- Vessel specific Anchor Handling Manual.
- Appendix 11 - B of the Guidelines.

Anchor handling operations will be executed under the direction of the PIC of the MOU or his delegate.

Exchange of information during anchor handling operations. The MOU is required to monitor the following and report any changes to the anchor handling vessels: disconnection and recovery of towing connections and assemblies; weather conditions and forecast weather window are suitable for operations; integrity of the unit, if relevant; MOU winch pay-out and recovery speeds; MOU propulsion assistance; communications between MOU cranes, winch operators, tow master and vessels. Vessels are required to monitor the following and report any significant changes to the MOU: international Collision regulations. Whilst working on MOU mooring equipment, alterations to heaving in or payout speeds and changes to vessel heading and power, should be made in a controlled manner. When vessels is heaving in or paying out moorings, reduce engine thrust if required to avoid damage. Be aware of mooring catenaries and required clearances. When anchor handling in adverse weather, dynamic forces are significant. Exercise great caution, monitoring environmental conditions continuously. Deployed and recovered equipment is to be visually checked for integrity and any deficiencies advised. Deviations to work scope and equipment used. Safe navigation whilst manoeuvring with anchors.

COMMUNICATIONS

Reference should be made to chapter 6 of the Guidelines relating to Communications. Communications in accordance with the Work Specification should be established and tested between the MOU work stations, vessel bridge and deck crews and between the various vessels involved in the operation. During anchor handling activities a common VHF radio channel should be designated for the use by the MOU and all vessels involved in the operation. This channel should not be used for other purposes whilst such activities are in progress. Where several vessels are working together on the same operation, a specific communication plan for that activity must be established which in particular ensures an effective and coordinated action in the event of any unintended incident. Communication between vessel workstations where the master and winch driver will be, and the anchor handling deck must be decided prior to the operation. Dependent on vessel's equipment and the operation concerned, the best means of communication may be personal UHF radios or by loudspeaker. Whichever means of communication is decided upon; it should be thoroughly tested prior to starting the operation.

VESSEL STABILITY

Stability of vessel is the responsibility of the master and should be checked prior to commencing operations with MOU. In addition to the sailing condition, stability calculations should consider worst case and predicted scenarios which may occur during and towards the end of a prolonged operations. This must take into account consumption of fuel and other consumables together with loading and deployment of chain or wire. These conditions are to be displayed and made available to MOU personnel on request throughout the operation and must be reviewed as soon as there is any event which may change the vessel's condition. Any specific conditions limiting the vessels stability (for example, the use and limits of stability tanks, minimum fuel requirements and free surface effect) are to be considered and readily available, and understood by all concerned. If stability computer systems are to be used these are to be verified and approved by Class or Flag. Prior to sailing, information must be readily available on the bridge, where it is visible to the navigator on duty, to show the acceptable vertical and horizontal transverse force / tensions to which the vessel can be exposed. This should show a GZ curve and a table of the tension and forces which give the maximum acceptable heeling moment. Calculations must show the maximum acceptable tension in wire or chain, including any transverse force, that can be accepted in order for the vessel's maximum heeling to be limited by one of the following angles:

- Heeling angle equivalent to a GZ value equal to 50% of GZ maximum.
- The angle of flooding of the work deck; i.e. the angle which results in water on working deck.
- A heel of 15 degrees.

Calculations should be made to show the maximum force from the wire or chain, acting down at the stern roller and transversely to the most extreme outer pins or stop in the event of no pins consider the outer edge of roller, which would be acceptable without taking the vessel beyond the angles stated above. The heeling moment based on transverse bollard pull must also be calculated and allowed for. The vertical component is to be taken as the distance (vertically) from the deck at the tow pins vertical centre line of thrust. If the calculations show that with predicted anchor chain or wire weights stability is outside the allowances stated above then it will be necessary to alter the vessels loading condition. Information available should also show the maximum force in the wire or chain as well as the point where the lateral force is assumed to be applied (towing pin or stern roller). The maximum vertical pull on the wire or chain must not be such as to exceed those limits given above nor to exceed the SWL of the roller. During deep water moves the weight on stern roller can be hundreds of tonnes, which may be applied at a distance off centre line according to the set-up of the towing pins. This may add to listing moments and stern trim, increasing the risk of a flooded deck, for example, from a breaking wave which can maybe result in a temporary reduction in stability. Normally, changes in the ballast condition should not carried out during anchor handling and towing operations, unless operations require a change in vessel condition. The effect of any such changes on the vessel stability must be fully evaluated and risk assessed. The status of all watertight, weather tight doors and hatches should maintained as set out in the vessel's operation manual. All such doors should be clearly signed to this effect. Further advice and guidance can be sought from:

- HSE Operations Notice Nos. 3, 6 and 65.
- HSE OSD 21 for Jack Ups.
- Warranty Certificate of Towage Approval.

GUIDANCE ON BOLLARD PULL AND VESSEL WORKING LIMITS

Vessel owners and masters should ensure that the vessel's bollard pull is adequate for the proposed operation. In considering this masters should be aware that bollard pull, as measured for the vessel's certificates in some cases does not allow for the power used by deck machinery, thrusters and other consumers diverted from the main propulsion. Allowance for any reduction should be made when considering bollard pull available during any operation. Maximum bollard pull is achieved with the cable right astern with rudders amidships A reduction in bollard pull must be allowed for should the angle of the cable lead other than right astern. Maximum tension whilst towing must never exceed 50% of MBL of weakest link in the assembly. It is recommended to aim for tension utilisation of 30% of the MBL to allow room for peak loads. Winch tension controls, where available, may be set with the above recommendations in mind. For anchor break out operations, the values above may need to be exceeded. This must be risk assessed and as agreed by all parties.

EMERGENCY RELEASE ARRANGEMENTS

To release excessive tension winches and mechanical stoppering devices are fitted with emergency release mechanisms. Maintenance and testing of these systems should be included as part of the planned maintenance regime and satisfactory operation should be verified before any anchor handling activities. The vessel's crew should be trained and competent in the operation of the emergency release systems in addition to being familiar with their reaction times and effect. Instructions giving information on how to operate the emergency winch stops and releases should be readily available on the vessel.

Further Guidance for Particular Anchor Handling Operations

Considerations for Deeper Water. Anchor handling operations in deeper water carry significant additional hazards and these may be location specific. In the context of this document "deeper water" is considered to be depths in excess of 300 metres (984 feet), though it is recognised that this is an arbitrary distinction, and any assessment of what might be considered as "deep water" operations will take into account the capabilities of the vessels supporting such activities. However, any operations beyond the continental shelf should always be considered as deep water activities. Where deep water anchor handling operations are being planned additional factors should be taken into account in addition to those involved in normal anchor handling operations: These include, but are not limited to, the suitability of vessel(s) for location specific operations taking into account environmental and other variables. To minimise damage to work wire from joining shackles use longer continuous lengths of work wire. Where joining is unavoidable to avoid damage use joining links rather than shackles. All wires to be spooled onto the drum under tension. Use work wire swivels to avoid twisting damage from the inherent high loads of deep water Anchor handling. Wires should be de-tensioned using suitable methods after use. Swivels are used to avoid twisting damage in the wire when exposed to high tension. Swivels are also used to reduce the risk of torsion building up in the wire. Such torsion could be released when disconnecting work wire from PCP, with risk of serious injury to personnel. Use of PCP chain tail in the shark jaws is particularly important in deep-water anchor handling operations. Buoys should not be "free launched" from deck but deployed in controlled manner to avoid shock load damage. Whilst deploying chain there may be a requirement for high tension to be used. Chain contact with the gypsy should therefore be maximised to avoid potentially dangerous slippage. In some circumstances, and where equipment is suitable, both gypsies may be used. Dynamic braking or tension control arrangements should be used, if available. Fibre ropes are frequently used in deep water mooring arrangements for a variety of different reasons. The original equipment manufacturers' guidelines for the use and handling of these products should always be complied with.

Breaking Out of Anchors. There is a possibility of damaging the PCP system or the work wire if the wire is overloaded whilst breaking the anchor loose from the bottom. It is unsafe to shorten up on the work wire (i.e. heave in on the vessel winch, until the vessel stern roller is vertically above the anchor position) in an attempt to break out an anchor. Doing so is liable to cause the wire, the winch, the anchor or other equipment to fail. Refer to *Figure 9-1*. The tension, which during the above mentioned method is used on the wire, is dependent on following circumstances:

- Winch pull force, depending on the size of the winch and how many layers are on the drum. If using one of the bigger winch sizes it is easily possible to exceed the breaking load of the wire.
- The vessel buoyancy combined with sea state forces in a sea will exceed the breaking load of the wire by many times regardless of the capacity of the winch.
- The diagram shows the relative vector forces for anchor removal with relation to vessel position and water depth.

In this illustration *50* is the minimum break loose force and optimum direction. The vectors show the significantly increased force required as the vessel nears the vertical pull. Forces at the anchor can be calculated using the formulae below:

$$LF = BP / \cos(\arcsin(WD / WL))$$

and

$$UL = BP \times \tan(\arcsin(WD / WL))$$

where

$$LF = \text{Line force in work wire}$$

- UL = Uplift force at anchor.
- BP = Bollard Pull being exerted.
- WD – Water Depth.
- WL = Wire Length deployed, measured from Stern Roller.

Anchors in very soft clay can be buried very deep, penetrations in excess of 100 metres (328 feet) being recorded is soft, estuary soils. The material behind the anchor will be disturbed as it penetrates the seabed. The best practice is therefore to find the optimum pull position and avoid overloading the equipment which would result in an unsafe situation. If problems are experienced in breaking out the anchor consideration should be given to using another vessel to assist, but any such assistance should be fully risk assessed prior to commencement of operations.

Figure 9-1 Relative Vector Forces in Anchor Recovery

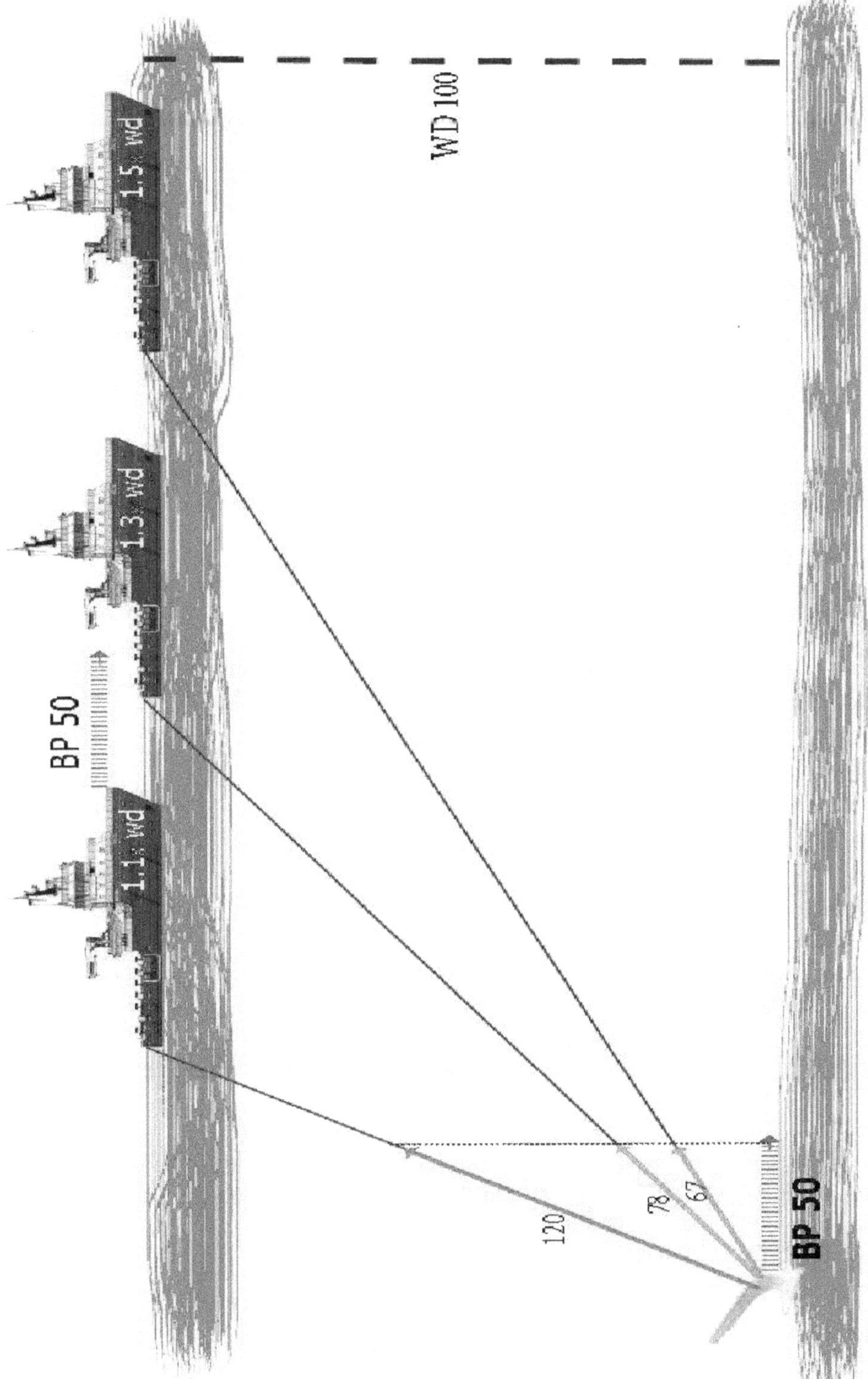

END NOTES

1 Significant sea height, wind and current velocities to be determined by operational considerations or risk assessment.

Chapter 10

Project Support Operations

Offshore operations require vessels to support a wide range of activities, including logistics support and supply; MOU moving, including anchor handling and towing; marine transportation; lifting operations, both above and below sea surface; and other operations, including well servicing and maintenance, diving and ROV support, IRM support, survey support, dredging and rock dumping; as well as emergency response and rescue support; and guarding and monitoring support. In general relate to the management and operation of all vessels approaching or operating in the vicinity of any offshore facility, regardless of whether a formal safety zone has been established. Chapters 9, 10 and 11 of the Guidelines relate particularly to operations included in Items 1 and 2 in the list. Other operations, including the use of smaller craft in supporting offshore operations, are considered in the remainder of this chapter.

Operations included. Marine transportations to which this section of the Guidelines relates may include, but are not limited to, transportation of unusual items on the deck of own vessel, particularly where the item involved is large, heavy or, if damaged, would result in significant actual or consequential loss; towage of a cargo barge onto which such items have been loaded and secured; and towage of any other vessel or floating object.

Operations excluded. It is not anticipated that this section of the Guidelines will relate to the following operations: logistics operations associated with the normal delivery or return of containers or similar cargo items to or from an offshore facility; and operations associated with moving a MOU from one offshore location to another, or to or from port facilities.

TOWING OPERATIONS

Passage planning and navigation. The master of the towing vessel is responsible for the preparation of a detailed passage plan and the subsequent safe navigation of the tow. Where appropriate, navigation warnings shall be broadcast by the tug at regular intervals. Where more than one vessel is utilised to tow the unit a lead tug will be nominated by the PIC of the operation, which may be a tow master if present. The master of this vessel will assume the responsibilities described above and shall also ensure that the other vessel(s) involved comply with the plans. This does not, however, relieve the master of any vessel from the responsibility of safeguarding the safety of personnel and equipment on board own ship. The passage plan must be carefully developed with regard to water depth, other offshore and subsea facilities, and emergency locations or refuges which may be utilised if required. Close attention should be paid to the length and catenaries of the tow wire and its relation to environmental conditions, water depth and vertical clearance over any sub-sea assets in the vicinity of any location or whilst on passage. Route must keep safe distance from any other facilities. Pass on the side that best assures tow will drift away from the facility in case of power loss or loss of tow. The passage plan shall not use facilities as way points. Regular weather forecasts should be provided. Normally two forecasts, twice per day, each prepared independently, are required. Communication lines as agreed during any pre-operational meetings should be observed.

Contingency and Emergency Towing Arrangements. Retrieval arrangements for the recovery of the towed object's main towing gear in the event of its failure should be fully operational. The towed object's emergency towing system should be rigged and ready for immediate use. Arrangements for this system to be recovered by a towing vessel in all weather conditions without the necessity of boarding the tow should be deployed. Where the mobilisation of any additional equipment to facilitate this is required, all such equipment should be readily available on board and checked to ensure it is fully operational.

Record of Towing Operations. Vessels engaged in any towing operations should maintain complete records of such activities. These records should normally consist of two parts, the Daily Log and the Voyage Summary. *Daily Log.* Information should be recorded at regular intervals whilst actually engaged in the towing operation, as follows:

- Date and time of log entry.
- Name of towed object.
- Position, actual or estimated.
- Power setting(s) on main propellers or thrusters.
- Length of tow-line deployed.
- Relevant weather conditions, for example particulars of wind, sea and swell.
- Any changes to towing configuration in foregoing period.

This information should be recorded more frequently in periods of severe weather. *Voyage Summary.* On completion of each towing operation a voyage summary should be prepared. This summary should include the following information:

- Particulars of towed object, including name and description.
- Date and place of commencement of tow.
- Date and place of completion of tow.
- Distance.
- Average speed throughout tow.
- Brief description of towing arrangements used.
- Brief description of weather experienced during tow.
- Brief description of any relevant incidents occurring during tow.

LIFTING OPERATIONS (ABOVE AND BELOW SEA SURFACE)

Operations included. Lifting operations to which this section of the Guidelines relates may include, but are not limited to, installation operations, involving use of equipment installed on vessel to lift items of equipment from deck of own or other vessel or barge for installation on an offshore facility; installation operations, involving use of equipment installed on vessel to lift items of equipment from deck of own or other vessel or barge for installation as part of a sub-surface facility; removal or de-commissioning operations, involving use of equipment installed on the vessel to lift items of equipment from an offshore facility onto deck of own or other vessel or barge; removal or de-commissioning operations, involving use of equipment installed on the vessel to lift items of equipment from a sub-surface facility onto deck of own or other vessel or barge; and any other operations involving use of the equipment installed on the vessel to support project related activities on an offshore facility, either above of below the sea surface, or on another vessel.

Excluded Operations. It is not anticipated that this section of the Guidelines will relate to logistics operations associated with the normal delivery or return of containers or similar cargo items to or from an offshore facility; lifting activities which may be required during MOU operations and "internal" lifting operations on the deck of own vessel, except where the item to be lifted is unusually large, heavy or, if damaged, would result in significant actual or consequential loss.

Particular Requirements for Sub-Surface Lifting Operations. In some areas of the world requirements relating to lifting operations below the surface of the sea may be different to those involved where the activities are undertaken only in air. Such differences may include specifications of equipment involved, the competencies of personnel involved, and operational planning and execution. It is the responsibility of all parties involved to ascertain the requirements for undertaking subsurface lifting activities in the current area of operations and ensure that these are complied with.

SPECIFIC OPERATIONAL PROCEDURES

It is likely that the planning and execution of operations to which sections 12.2 and 12.3 of the Guidelines relate will require the development of specific procedures describing the activities involved. Where relevant, these procedures will be based on Engineering design and analysis, Hazard identification and risk assessments. Where relevant, combined operations safety cases, including relevant bridging documents and simultaneous operations assessments and reviews. Each set of such procedures will be developed by the relevant project team specifically for the operations being contemplated, and will be subject to the relevant quality assurance and approval processes. Owing to their specific nature such documents are generally developed for the particular project involved. Further detailed consideration of the development of these procedures in therefore beyond the scope of this publication.

GENERAL GUIDANCE

Whilst further consideration of the specific operational procedures are outside the scope of this publication. Further information relating to particular aspects of the planning and execution of project-related on- and offshore operations can be obtained from a variety of sources, some of which are listed in *Table 10-1*. The documents included in *Table 10-1* are regularly reviewed and revised by the publishers. The latest revisions, which, with the exception of the document published by DNV, are available on the internet web sites of the relevant organisations should always be referred to when planning or executing any of the operations to which they relate.

Table 10-1 Further General Guidance

Source	Document Reference	Document Particulars
		Title
DNV	13	Rules for Planning and Execution of Marine Operations
GL Noble Denton	21	Guidelines for Load Outs

GL Noble Denton	27	Approval of Towing Vessels
GL Noble Denton	30	Marine Lifting Operations
IMCA	M187	Guidelines for Marine Transportations
IMCA	M171	Lifting Operations
IMCA	M193	Communications during Lifting Operations
IMCA	M194	Wire Rope Integrity Management for Vessels in Offshore Industry
IMCA	M203	Simultaneous Operations
IMO	MSC 494	Safety of towed ships and other floating objects

MARINE WARRANTY SURVEYOR INVOLVEMENT

In many instances the planning and execution of operations to which these sections of the Guidelines relate will require the approval of an accredited marine warranty survey practitioner (MWS). The MWS will review the proposed procedures and revert with relevant comments and recommendations. Approval will often be subject to the attendance on site of the MWS's representative, who may communicate further recommendations as thought fit to the representative of the insured party on whose behalf he is acting. Compliance with the MWS's recommendations, or agreed alternatives, is strongly recommended since failure to do so may compromise the insured party's commitments to its insurance underwriters and in the event of an incident resulting in loss may expose it to significant commercial risk. Where the services of a marine warranty surveyor have been retained by any insured party the role and expectations should be advised to all other parties involved in the relevant operations.

SUPPORT FOR OTHER OPERATIONS

The planning and execution of any operations referred to in section 12.1(5) of the Guidelines which supported by offshore vessels will involve the development of task-specific procedures for the work to be undertaken. The contents of such procedures are outside the scope of this publication. However, when any vessel supporting such operations is approaching or operating in the vicinity of any offshore facility, irrespective of whether a formal safety zone around it has been established, all relevant recommendations included in should be observed. A note to this effect should be included in the procedures referred to above.

RESPONSE AND RESCUE SUPPORT

Primary functions. At many offshore facilities elements of the arrangements for emergencies which the operator must establish to ensure the safety of the workforce on board in the event of an incident is supported by vessels mobilised to provide response and rescue services. The primary functions of vessels mobilised for this purpose include the following:

- Rescuing personnel who have inadvertently entered the water in the vicinity of an offshore facility and providing suitable facilities for their subsequent care.
- Monitoring the movements of other marine traffic in the vicinity of the facility and taking appropriate action where risk of collision with it is though to exist.
- Acting as contingency command and control station should an incident on the facility result in its own arrangements being disabled.

Except otherwise advised these functions are to be supported on a continuous basis and no other activities should be undertaken by or on board the vessel which would compromise its ability to do so. In the event of any incident on the vessel which may result in it not being able to fulfil any of the functions above the management of the facility which it is supporting must be immediately be advised of this fact in order that appropriate alternative arrangements can be made.

General requirements. Vessels mobilised to provide emergency response and rescue services should comply with the requirements relating to such support which exist in the jurisdiction in which they are operating. It is the responsibility of all parties involved to ascertain the require ments for vessels providing support of this nature in the current area of operations and ensure that these are complied with.

Table 10-2 Response and Rescue Support Guidance

| Source | Document Particulars | |
	Document Reference	Title
DMA		Guidelines for Standby Vessels
ERRVA		Emergency Response and Rescue Vessels Survey Guidelines
ERRVA		Emergency Response and Rescue Vessels Management Guidelines
NOGEPA	6 and 7	Safety Standby Vessels Code of Practice
NOROGA		Reference to include

Where no such requirements exist, the documents listed in *Table 10-2* may provide useful guidance in ensuring that adequate standards for response and rescue support are established and maintained.

Response criteria. The jurisdiction under whose regime the facility is operating may establish criteria relating to time taken to rescue personnel from the water and transfer them to a place of safety in a variety of incident scenarios. Alternatively, it may require the operator of the facility to establish such criteria. It is in the interests of both the charterer and owner of the vessel to ensure that such criteria can be complied with. Facilities and equipment should therefore be provided for the on-going training of the personnel involved. Exercises to ensure that personnel remain familiar with the equipment and procedures involved should be arranged at frequent intervals. Where possible such exercises should be undertaken in typical environmental conditions which may be experienced at the location, always on the understanding that personnel or equipment should not be subject to unnecessary risk. Full records of any exercises undertaken should be retained for subsequent inspection by interested parties. From time to time there may also be a requirement for such exercises to be observed and recorded by an independent witness.

Adverse weather criteria. Facility operations which can be supported by a typical emergency response and rescue vessel will be dependent on the prevailing environmental conditions. The criteria under which the various levels of support can be provided will normally be agreed on an industry-wide basis for the area in which the facility is located. Where this is not the case, such criteria should be agreed between the charterer and the owner, in consultation with experienced masters, at the time of taking the vessel on hire. Typical adverse weather criteria for a conventional SBV, equipped with FRC's and or daughter craft for rescuing personnel from the water in lower sea states and a mechanical means or recovery in the more severe conditions when rescue craft cannot safely be deployed or recovered are included in *Annex E*. It should be noted, however, that in a long period, regular swell rescue craft may be safely deployed and recovered in higher sea states than indicated. On the other hand it may be that such craft cannot be safely used where waves are of short period and or confused at lesser heights than those indicated. In such circumstances other means of recovery should be considered. Furthermore, familiarity with the equipment and techniques involved in the rescue of personnel from the sea gained from suitable training together with frequent exercises in more challenging conditions may well enable the criteria set out in the Appendix to be extended. In addition, other technologies are available which may not be subject to the same environmental limitations. The master of the SBV will decide which equipment and techniques are most appropriate for use, based on circumstances associated with the emergency itself and the prevailing environmental conditions.

SBV operational capability. Section 7.2 of the Guidelines relates to the factors to be taken into account when assessing the operational capability of any vessel, including those providing response and rescue support. However, with particular regard to vessels providing such support it should also be borne in mind that the vessel is effectively an integral part of the facility's emergency response arrangements. A reduction in its operational capability, particularly where this involves loss of manoeuvrability or ability to deploy and recover its rescue equipment, may therefore compromise any commitments to the relevant authorities or the workforce made by the facility operator to maintain an acceptable standard of emergency response arrangements. Any reduction in the operational capability of a vessel providing response and rescue support at an offshore facility should therefore be notified to the manager at the earliest opportunity in order that suitable alternative arrangements can be established without delay.

Weather side working. Sections 7.3.1 and 8.11 of the Guidelines relates to the factors to be taken into account when assessing setting up and working on the weather side of an offshore facility and relates also to vessels providing response and rescue support. In assessing whether the vessel can provide effective support on the weather side of the facility the master should bear in mind any potential impact on the deployment of rescue equipment. In general, any request to take up station directly up-wind or up-current of the facility should be challenged, since in most instances equally effective support can be provided from a position where the vessel has more freedom to manoeuvre and is not in a "drift-on" situation.

Work parties outside the perimeter of the facility. From time to time it may be necessary for work to be undertaken on the facility outside its normal barrier perimeter, which may include any scaffolding assembled and maintained in accordance with the relevant rules. Such work may be referred to by a variety of terms, including "over side", "outboard", "overboard", etc. Such work generally involves an increased risk of personnel entering the water. Therefore, when in progress a higher state of readiness should be maintained on the vessel providing response and rescue support. This may be referred to as providing "close stand-by" support. When requested to provide such support the master should take the following actions:

- Establish details of personnel at risk, including numbers and locations.
- Ensure that personnel and equipment on the vessel are at the required state of readiness.
- Ensure that the vessel is maintained in a position relative to the facility and the environment such that rescue facilities can be deployed in the most expeditious manner.
- Ensure that the terminology to be used has been agreed and understood by all involved.
- Ensure that communication have been established and are maintained with the watchmen responsible for monitoring the activities of each work party.

It is not the responsibility of the vessel to maintain a visual watch of the various work-sites. Any request to maintain such a watch should be challenged, since this could compromise the safe navigation of the vessel and would be impossible where several work-sites are involved.

Sharing of SBV support services. Where several offshore facilities are located within close proximity of each other the services of one vessel may be shared between them. This may include situations where several operators are also involved. In such circumstances clear procedures and protocols should be developed and agreed between all the parties involved, including the vessel owner, to ensure that the primary functions referred to above can be supported at all the facilities involved. For clarity the agreed procedures and protocols should be consolidated into a vessel sharing manual, contents of which should include as a minimum:

- Details of the parties involved.
- Details of ownership of manual, control and distribution.
- Details of facilities involved in the vessel sharing.
- Organograms for each facility, showing relationship with vessel.
- Details of where vessel will take up station in various circumstances.
- Activities which can be supported at each location, dependent on vessel's position, environmental conditions or other circumstances. In particular the following should be clearly

described:

  - Operations which can be supported simultaneously at all facilities supported.
  - Operations which require the vessel to take up station at a particular location.
  - In the case of the above, operations which then could not be supported at any facility.

- Coordination of vessel operations and decision making process.
- Arrangements for monitoring movements of other marine traffic and in particular actions to be taken where threat of collision involving any of the facilities support is thought to exist.
- Arrangements in event of equipment failure on vessel.
- Arrangement for regular vessel relief.

Supplementary information may be included in appendices to the manual, including:

- Contact details for each party involved (if not included in main document).
- Further information, charts or data sheets relating to each of the facilities supported, general field information.
- Details of the vessel(s) involved.
- Further information relating to vessel's normal patrol zone, including response times at each facility where relevant.
- Information to assist with planning vessel movements.
- Response criteria for each party involved.
- Adverse weather policies for each party involved.

Other support functions. In some instances vessels mobilised to provide response and rescue services may be constructed, outfitted, equipped and manned to support other functions. Such functions may include:

- Cargo carrying, both deck and bulk goods.
- Fire fighting capability.
- Oil recovery and pollution prevention.

Additional activities associated with other such functions may be undertaken simultaneously with providing such services provided that the vessel's ability to fully support the primary roles referred to above are never compromised. The planning of and procedures describing any operation requiring the support of a vessel which is simultaneously providing emergency response and rescue support should always recognise this requirement. Where a vessel is supporting any additional activities whilst simultaneously providing response and rescue support the relevant recommendations in relating to such activities should be observed.

GUARD CHASE VESSELS

General requirements. From time to time a requirement may be identified for a vessel to be chartered to maintain watch over an asset or equipment associated with offshore operations. Such a requirement includes warning other vessels in the vicinity should their activities or actions be thought to pose a risk to facilities involved. Such vessels may include:

- Guard vessels chartered for the purpose of protecting and monitoring marine traffic in the vicinity of fixed assets of any type.
- Chase vessels chartered for the purpose of protecting the array of streamers deployed in the course of seismic survey operations.
- Any other vessels chartered for similar purposes.

Vessels mobilised for this purpose should comply with the requirements relating to such support which exist in the jurisdiction in which they are operating. It is the responsibility of all parties involved to ascertain the requirements for vessels providing support of this nature in the current area of operations and ensure that these are complied with. Where no such requirements exist the documents listed in *Table 10-3* may provide useful guidance in ensuring that adequate standards for this support are established and maintained.

Other requirements. In addition to the requirements relating to the outfitting and equipping of vessels set out in the above documents it is also recommended that, if not required by legislation, where a single watch keeper is likely to be on duty for extended periods, a watch alarm should also be fitted. This alarm should be fitted with the following features:

- Positive response by watchkeeper required at periods not exceeding 15 minutes.
- Response arrangements to be located remotely from any seating facilities.
- In event of non-response from watchkeeper general alarm to be activated throughout accommodation within further five minutes.
- Override or disablement of any of the above facilities only being possible with positive agreement of master and chief engineer.

Table 10-3 Guard and Chase Vessel Guidance

Source	Document Particulars	
	Document Reference	Title
ERRVA		Emergency Response and Rescue Vessels Survey Guidelines
ERRVA		Emergency Response and Rescue Vessels Management Guidelines
NOGEPA	6 and 7	Safety Standby Vessels Code of Practice
SFF, NFFO		Guard Vessel Good Practice

Further information to be provided to masters and skippers. Where vessels are taken on hire to support any asset protection, guarding or monitoring requirements the following information should be provided:

- Full particulars of the assets to be protected, guarded or monitored, including plans of any surface and or sub-sea architecture.
- Reporting protocols and relevant contact arrangements.
- Actions to be taken in the event that an approaching vessel is assessed as posing a threat to the assets being guarded or monitored.

This information should also be provided to any ERRV which may be required to guard or monitor assets in the vicinity of the offshore facility being supported.

Operational categories. It should be noted that some of the vessels to which this section of the Guidelines relates may be operated under the provisions relating to small water craft referred to below.

Small vessels and water craft. From time to time operations either off- or near-shore may be supported by smaller vessels or water craft.

Vessel or craft included. Vessels or craft involved may include:

- Any vessel or craft with load water line length of 24 metres (78 feet) or less.
- Vessels or craft with displacement hull forms within this length range.
- Rigid or semi-rigid hull forms within this length range.

Offshore, such craft may be deployed from a larger host vessel, though in certain circumstances they may be capable of autonomous operation. For near-shore operations they are likely to operate independently from convenient port or other safe haven.

Vessels or craft excluded. This section of the Guidelines does not relate to the following craft:

- Fast rescue boats mobilised on any vessel in compliance with the International Life Saving Appliance Code, as modified from time to time.
- Fast rescue craft mobilised on any Emergency Response and Rescue Vessel.
- Daughter craft mobilised on any Emergency Response and Rescue Vessel to provide extended rescue and response support which are operated under the provisions of a Loadline Exemption or similar arrangements.
- Small work-boats deployed from a larger vessel to support its operations, for example buoy boats associated with the maintenance of seismic cable arrays.

Craft of this type are likely to operate under provisions attached to the vessel from which they are deployed or dispensation from the relevant Flag State.

Construction, Operational and Competency Requirements. Small craft mobilised to support either off- or near-shore operations should comply with the requirements relating to such support which exist in the jurisdiction in which they are operating. It is the responsibility of all parties involved to ascertain the requirements for such craft in the current area of operations and ensure that these are complied with. Where no such requirements exist the documents listed in *Table 10-4* may provide useful guidance in ensuring that adequate standards are established and maintained. Other guidance exists, but that prepared by the MCA which is referred to above is unique in that matters relating to construction, operation and competencies are addressed in a single document. It should also be noted that where small craft are mobilised to provide response and rescue support additional requirements are likely to exist.

Further Information to be provided to the skipper or coxswain. Where small craft are taken on hire to support any operations the following information should be provided:

- Full particulars of the nature of operations to be supported.
- Reporting protocols and arrangements.
- Actions to be taken in the event of an emergency or other unforeseen events.

Table 10-4 Small Vessel and Water Craft Guidance

Source	Document Reference	Document Particulars
		Title
MCA	MGN 280	Small Vessels in Commercial Use Code
OGP		Watercraft Guidelines

Chapter 11

Emergencies

Facility operators and vessel owners shall prepare and maintain emergency preparedness procedures for those assets for which they are responsible. The operating company shall prepare and provide to all parties a bridging document linking the involved parties' emergency procedures. This shall include but not be limited to organisation charts, relevant communication, and contact numbers, notification plans, area charts showing facilities, etc., an overview of local resources in emergency: vessels, helicopters, etc., incidence response protocols, and media response protocols.

FACILITY EMERGENCIES

When facility emergency alarms sound OIM / facility manager will issue instructions to all vessels. Any emergency is to be handled in accordance with the Emergency Preparedness Procedures.

VESSEL EMERGENCIES

Within the Safety Zone, vessel emergency to be handled in accordance with the Emergency Preparedness Procedures. OIM or facility manager shall be informed to the nature of the emergency. Upon being notified of an emergency on any vessel within the Safety Zone he should take such actions as necessary to minimise the risk to the facility and its personnel. Outside the Safety Zone emergencies on vessels shall be handled in accordance with the owner's Emergency Response Plan, as described in their Safety Management System. Local and Regional agencies will be involved as necessary.

PORT EMERGENCIES

The ship-owner and the vessel's master are responsible for ensuring provision of adequate internal emergency procedure covering their own vessel as well as ensuring sufficient familiarisation with relevant procedures from Port Authority and base company in this respect.

Chapter 12

Further Information and References

Further relevant material relating to the information included in this guidance document is available from a variety of sources, a number of which are listed in *Table 12-1*. Where possible, this information is open source, available in the public domain and can be viewed at or downloaded from the source's web portal, which are also listed as hyper-links.

NOTES

- Reference numbers only included if the source is specifically referred to in this publication. Sources are categorised by the originator or sponsor of the relevant information.
- Application is categorised as follows:

    - 1 = International, applicable in all operating areas.
    - 2 = General, may be used if no regional or flag guidance is available.
    - 3 = Regional or flag guidance, usually relating particularly to the country of origin
    - 4 = For information only.

- The latest revision of all such sources posted on the relevant web portal should always be referred to.

Table 12-1 Further Information and References

Document Particulars			Country of Origin	Application	Available From
Nos	**Title**	**Originator Sponsor**			
	Rules for Planning and Execution of Marine Operations	DNV	NO	1	Not available electronically
	Emergency Response and Rescue Vessels, Management Guidelines	ERRVA and OGUK	UK	2, 3	www.marinesafety-forum.org/
	Emergency Response and Rescue Vessels, Management Guidelines	ERRVA and OGUK	UK	3	www.marinesafety-forum.org/
0013	Guidelines for Load Outs	GLND	UK	1	www.gl-nobleden-ton.com/en/rules_guidelines.php
0021	Guidelines for the Approval of Towing Vessels	GLND	UK	1	www.gl-nobleden-ton.com/en/rules_guidelines.php
0027	Marine Lifting Operations	GLND	UK	1	www.gl-nobleden-ton.com/en/rules_guidelines.php
0028	Guidelines for Transportation and Installation of Steel Jackets	GLND	UK	1	www.gl-nobleden-ton.com/en/rules_guidelines.php
0030	Guidelines for Marine Transportations	GLND	UK	1	www.gl-nobleden-ton.com/en/rules_guidelines.php
0032	Guidelines for Moorings (In course of preparation)	GLND	UK	1	www.gl-nobleden-ton.com/en/rules_guidelines.php
MGN 13-2	Location Moves of Self Elevating Units	GLND	UK	1	www.gl-nobleden-ton.com/en/rules_guidelines.php
MGN 14-2	Location Moves of Semi Submersible Units	GLND	UK	1	www.gl-nobleden-ton.com/en/rules_guidelines.php
	Role of the Warranty Surveyor	GLND	UK	4	www.gl-nobleden-ton.com/en/rules_guidelines.php
	Guidelines for Pipeline Operators on Pipeline Anchor Hazards	HSE	UK	2,3	www.hse.gov.uk/offshore/index.htm
	Guidelines to HSR in Britain	HSE	UK	3	www.hse.gov.uk/offshore/index.htm
	LOLER - Guidance	HSE	UK	3	www.hse.gov.uk/offshore/index.htm
	LOLER 1998 (Open Learning)	HSE	UK	3	www.hse.gov.uk/offshore/index.htm
MaTSURR 307	Use and Operation of Daughter Craft in the UKCS	HSE	UK	3	www.hse.gov.uk/offshore/index.htm

OIS 1	Performance of SBV Radar	HSE	UK	2	www.hse.gov.uk/offshore/index.htm
OTO 1997 - 058	Good Practice in Offshore Rescue	HSE	UK	2	www.hse.gov.uk/offshore/index.htm
OTO 2001 - 040	Guidance on Personnel Transfer by Carriers	HSE	UK	2,3	www.hse.gov.uk/offshore/index.htm
OTO 9906a	Effects of Weather on Performance Times in Offshore Rescue (1)	HSE	UK	2	www.hse.gov.uk/offshore/index.htm
OTO 9906b	Effects of Weather on Performance Times in Offshore Rescue (2)	HSE	UK	2	www.hse.gov.uk/offshore/index.htm
RR 049	Safety of Jack-Up Units in Transit	HSE	UK	2	www.hse.gov.uk/offshore/index.htm
RR 108	Performance Capabilities of Daughter Craft Crews	HSE	UK	2	www.hse.gov.uk/offshore/index.htm
RR 569	Analysis of ERRV Trials	HSE	UK	2	www.hse.gov.uk/offshore/index.htm
	MCA and MAIB Memorandum of Understanding	HSE	UK	3	www.hse.gov.uk/offshore/index.htm
D 035	The Selection of Vessels of Opportunity for Diving Operations	IMCA	UK	1	www.imca-int.com/
M103	Guidelines for The Design and Operations of Dynamically Positioned Vessels	IMCA	UK	1	www.imca-int.com/
M109	A Guide to Dynamic Position Related Documentation of Dynamically Positioned Vessels	IMCA	UK	1	www.imca-int.com/
M117	The Training and Experience of Key Dynamic Positioning Personnel	IMCA	UK	1	www.imca-int.com/
M125	Safety Interface Document for Dynamically Positioned Vessel Working Near an Offshore Installation	IMCA	UK	1	www.imca-int.com/
M125a	Check List associated with M125	IMCA	UK	1	www.imca-int.com/
M131	Review of Use of Fan Beam Systems for Dynamic Positioning	IMCA	UK	1	www.imca-int.com/
M139	Standard Report for DP Vessels Annual Trials	IMCA	UK	1	www.imca-int.com/
M141	Guidelines for the Use of DGPS as Position Reference in DP Control Systems	IMCA	UK	1	www.imca-int.com/
M149	Common Marine Inspection Format	IMCA	UK	1	www.imca-int.com/

M163	Guidelines for Quality Assurance and Quality Control of Software	IMCA	UK	1	www.imca-int.com/
M171	Crane Specification Document	IMCA	UK	1	www.imca-int.com/
M182	International Guidelines for the Safe Operation of Dynamically Positioned Offshore Supply Vessels	IMCA	UK	1	www.imca-int.com/
M187	Lifting Operations	IMCA	UK	1	www.imca-int.com/
M189	Inspection Checklist for Small Work boats	IMCA	UK	1	www.imca-int.com/
M193	Guidelines for Operational Communications - Lifting Operations	IMCA	UK	1	www.imca-int.com/
M194	Wire Rope Integrity Management for Vessels in Offshore Industry	IMCA	UK	1	www.imca-int.com/
M202	Guidelines for the Transfer of Personnel to or from Offshore Vessels	IMCA	UK	1	www.imca-int.com/
M203	Simultaneous Operations	IMCA	UK	1	www.imca-int.com/
A 893, Annex 25	Guidelines for Passage Planning	IMO		1	
MSC 494	Safety of Towed Ships and Other Floating Objects	IMO		1	
MSC 645	Guidelines for Vessels with Dynamic Positioning Systems	IMO		1	
	A Masters Guide to the UK Flag	MCA	UK	2,3	www.dft.gov.uk/ mca/mcga07-home/ shipsandcargoes/ mcga-shipsregsand-guidance.htm
MGN 280 - 3(M)	Code - Small Vessels in Commercial Use	MCA	UK	2	www.dft.gov.uk/ mca/mcga07-home/ shipsandcargoes/ mcga-shipsregsand-guidance.htm
MGN 282	Dangerous Goods - Guidance in the Carriage of Packaged Dangerous Goods on Offshore Supply Vessels	MCA	UK	2,3	www.dft.gov.uk/ mca/mcga07-home/ shipsandcargoes/ mcga-shipsregsand-guidance.htm
MGN 283	Dangerous Goods - Guidance on the Back-loading of Contaminated Bulk Liquids from Offshore Installations to Offshore Supply / Support Vessels	MCA	UK	2	www.dft.gov.uk/ mca/mcga07-home/ shipsandcargoes/ mcga-shipsregsand-guidance.htm

MGN XXX	Ship to Ship Transfers	MCA	UK	2	www.dft.gov.uk/ mca/mcga07-home/ shipsandcargoes/ mcga-shipsregsand-guidance.htm
MSCP 01	Code of Safe Working Practices for Merchant Seamen	MCA	UK	2,3	www.dft.gov.uk/ mca/mcga07-home/ shipsandcargoes/ mcga-shipsregsand-guidance.htm
MSN 1767	Hours of Work, Safe Manning and Watch Keeping	MCA	UK	2,3	www.dft.gov.uk/ mca/mcga07-home/ shipsandcargoes/ mcga-shipsregsand-guidance.htm
	RNLI Rescue Boat Code	MCA	UK	2,3	www.dft.gov.uk/ mca/mcga07-home/ shipsandcargoes/ mcga-shipsregsand-guidance.htm
	Anchor handling vessel, PSV HSE Inspection Check List	MSF	UK	1	www.marinesafety-forum.org/
	Anchor Handling Manual Template	MSF	UK	1	www.marinesafety-forum.org/
	Guidelines for Contents of MOU Move Procedures	MSF	UK	1	www.marinesafety-forum.org/
	SF - 06.32 Cherry Picking	MSF	UK	2	www.marinesafety-forum.org/
	SF - 07.29 Ships Access in Harbour	MSF	UK	2,3	www.marinesafety-forum.org/
	DP Operations Guide	MTS	US	1	www.mtsociety.org/
	Safety Measures for Anchor Handling Vessels and Mobile Offshore Units	NMD	NO	2,3	
IGC	Safety Standby Vessels	NOGEPA	NL	2,3	www.nogepa.nl/ en/Home/Olie-gasinNederland/ Veiligheidgezond-heid/Industrierichtlij-nen.aspx
OPS 013	Collision Risk Avoidance	O&GUK	UK	2,3	
	Watercraft Guidelines	OGP		2	
	Transfer of Personnel using Small Boats	PGS		2	
	Bulk Hose Handling Guidelines	STEP-CHANGE	UK	2	www.marinesafety-forum.org/
	Health and Safety Management Systems Interfacing	STEP-CHANGE	UK	2,3	

	Guidelines for Anchor Handling in Vicinity of UKCS Installations, Pipelines and their Sub Sea Equipment	UKOOA	UK	2,3	

Annexes

Annex A

Hierarchy of Authority

The hierarchy of authority for the guidelines set out in this publication within the overall maritime environment is summarised in *Figure A-1*.

INTERNATIONAL MARITIME ORGANISATION (IMO)

This is the United Nations body supported by all major maritime nations and responsible for international marine legislation which is then ratified by and included in the statute books of member countries. Such legislation is supported by four primary "pillars", as follows:

1. International Convention for the Prevention of Pollution from Ships (MARPOL).
2. International Convention for the Safety of Life at Sea (SOLAS).
3. International Convention for Standards of Training, Certification and Watchkeeping for Seafarers (STCW).
4. Maritime Labour Convention (MLC).

Pillars 1-3 have been ratified and are currently in force. Following ratification in 2012 Pillar 4 was implemented in 2013. Subsidiary legislation flows from each of the pillars referred to above.

FLAG STATE

In addition to international regulations flowing from the IMO as described above the state where the vessel is registered may have further, supplementary rules. These are likely to include those developed by Classification Societies acting on behalf of the flag state.

REGIONAL OR LOCAL AUTHORITIES

Compliance with legislation originating from IMO or the flag state is mandatory wherever the vessel is operating geographically. However, the authorities responsible for marine activities in its present trading area may have further local requirements relating to, for example, such matters as environmental measures, etc. It is the responsibility of vessel owners and masters to ensure that all the above requirements are understood and complied with at all times. This document has been prepared on the understanding that this is indeed the case.

CHARTERER'S SPECIFIC REQUIREMENTS

Each charterer may have specific requirements, both relating to general marine activities or of a more particular nature should a vessel be taken on hire for any specialised operation. Such requirements may be summarised in the charterer's marine operations manual or in any relevant project-specific procedures It is the responsibility of the charterer to ensure that owners and masters are made aware of any such specific requirements.

Fig.A-1 Hierarchy of GOMO Authorities

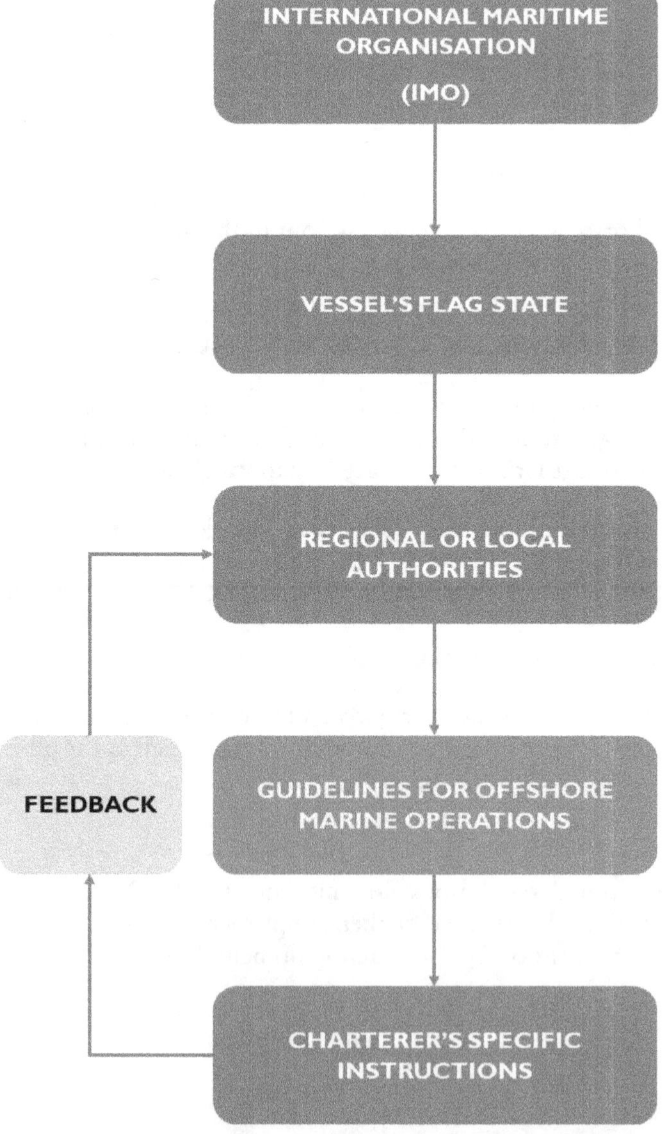

Example Platform and MOU Data Card

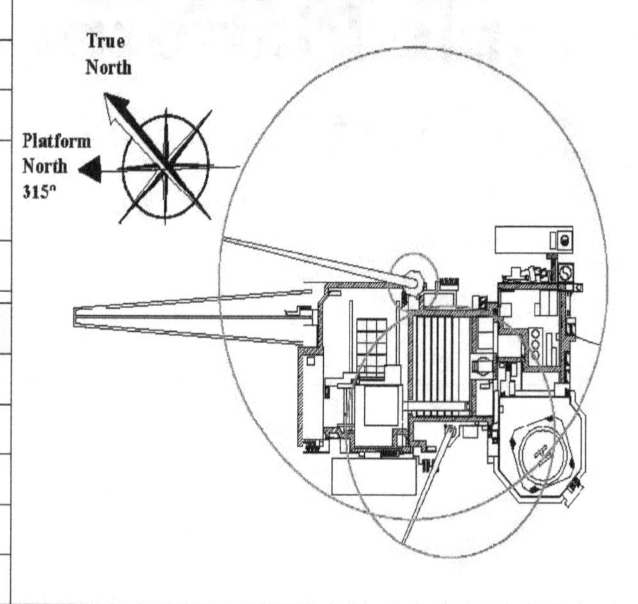

Chevron	**Alba North** **UKCS 16 / 26** **PSV / AHV Data Card**

Location	Alba Field 16 / 26
Latitude	58° 03' 31" N
Longitude	01° 04' 53" E
Heading	315° (T)
Water Depth	138 metres 453 feet
Call Sign	MPTK4

Specific Marine Hazards

- Various pipelines, umbilicals etc.
- Overboard discharges
- Field activities e.g. shuttle tanker
- Tidal information
- Installation ongoing operations

Communications	General	Emergency
VHF	Ch. 108 (P1) Ch. 50 (P2) Ch. 8 & 12	Ch. 16
Telephone	01224 334000	00 871 (874) 144 5734
Fax (Radio Room)	01224 335680	N/A

Helicopters	Bristows
Log Traffic Emergency	126.400 MHz 123.550 MHz 121.500 MHz
Telephone Tel. (out of hours)	01224 756214 01224 756321
Fax	01224 756348

East Crane	Approx. SWL	Radius	Sea SWH	West Crane
Whip Line	15 Tonne	23 m	1.6 m	The west crane is not normally used for Marine op's due to lack of visibility. Priority lifts that do not exceed 5 Tonnes at max. radius and min. sea state can be carried out with the permission of the OIM.
Whip Line	5.2 Tonne	40 m	3.9 m	
2 Fall	24 Tonne	18 m	1.6 m	
2 Fall	16.6 Tonne	22 m	2.8 m	
2 Fall	12.6 Tonne	40 m	3.9 m	

Nearby Installations			Shore Distances	
Alba FSU	(Chevron)	248° (T) x 1.6 mls.	Aberdeen	241° (T) x 115 mls.
Britannia	(Britannia)	108° (T) x 1.9 mls.	Peterhead	250° (T) x 97 mls.
Andrew	(BP)	094° (T) x 10.3 mls.	Wick	280° (T) x 134 mls
Balmoral	(AGIP)	005° (T) x 10.3 mls.	Sumburgh	326° (T) x 130 mls.

Rig Alarms	Fire & Emergency	Abandon Rig	Toxic Gas
Sound	Intermittent	Continuous Variable Tone	Continuous Steady Tone
Light	Flashing Yellow	Flashing Yellow	Flashing Red

Hoses & Connections   (East Crane)

Grade	Connection
Diesel	4" Avery Hardol
Pot Water	4" Weco
Drill Water	4" Weco
Cement	4" Weco
Barite / Bent.	4" Weco
Liquid Mud	4" Avery Hardol
Brine	4" Avery Hardol

Cargo Transfer Operations

- Agree product & quantity to be transferred
- Ensure agreed pump pressures / rates before discharge commences
- Confirm who will decide the routine halting of the operation
- Cargo transfer should commence slowly to check integrity of system
- Rates should only be increased when integrity of hose proved
- Confirm regularly with Control Room the amount of cargo transferred

Vessel Co-ordination

The co-ordination of Marine Operations will be as per the Master's Instructions issued to the vessel by Team Marine prior to departure from port. Daily reporting requirements are detailed within these instructions.

When in field area report :

Name of vessel
Arrival at the installation
Departure from the installation
Upon being sent to standby
Upon ceasing operations due to weather

Upon leaving the field report :

Name of vessel
Location and time of departure
ROB bulk products
Fuel and water requirements
ETA at port or next installation
Deck area utilisation (e.g. 80%)
Liquid tank status

Oct. 2005
Rev. 4

Example Base Operator and Port Data Card

PETERHEAD INFORMATION SHEET

BASE RULES
Vessel to be adequately manned at all times.

All personnel must wear appropriate PPE & High Viz at all times when outside of vessel's accommodation, including when transiting the base.

A safe means of access must be provided by vessel.

Moorings and Gangway to be properly tended.

It is the Master's responsibility to ensure that there is sufficient water depth under the keel.

Weather forecasts are available from on Ch 11.

The Master and Mates should read North Sea Section

PRIOR TO ARRIVAL AT
CONTACT PORT CONTROL ON CH 16 & CH 14.
CONTACT   BASE ON CH 11 1 HOUR PRIOR TO ARRIVAL AND AT 1 MILE FROM THE BREAKWATER.

Station	Frequency & Telephone No
* Harbours	Ch 16 & Ch 14
* Base	Ch 11

NAME	FUNCTION	CONTACT NO
*	Marine Operations Supervisor - In charge of the day to day running of the Supply Vessel Fleet.	* or through Operations Desk
*	Marine Co-ordinator - Responsible for scheduling of vessels. Main point of contact for vessels on charter to *	* or through Operations Desk
*	Assistant Marine Co-ordinator. Back-up to *	* or through Operations Desk
Operations Desk	Responsible for berthing of vessels, organising fuel, water, skips, labour, craneage. Boatmen. Main point of contact for vessels not chartered to *	* or through Operations Desk
*	Marine Technical Manager - Responsible for ensuring vessels are fit for purpose, incident/accident investigations, answering any queries of a technical nature.	* or through Operations Desk
*	Assistant Marine Technical Manager - Responsible for running the Stand By Vessel fleet and providing technical back up to the Marine Technical Manager.	* or through Operations Desk
*	* Department	* or through Operations Desk
*	* Surveyors, carry out tank inspections and verify liquid bulks loaded.	* or through Operations Desk
*	Environmental Chemist - will give advice on any chemicals being carried or requiring disposal.	*
*	Marine Operations Manager - Responsible for Anchor Handlers.	*

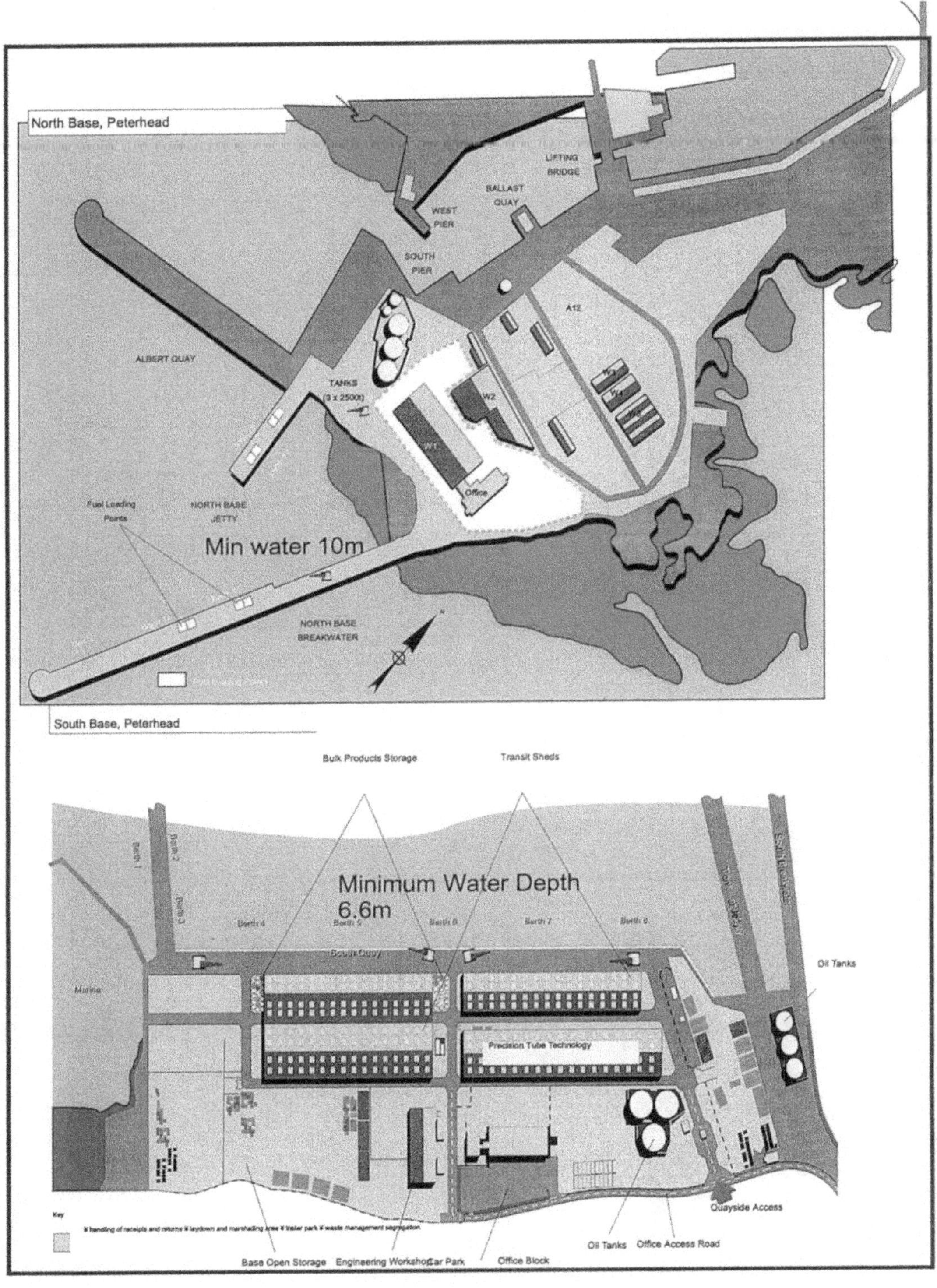
North Base, Peterhead
LIFTING
BRIDGE
BALLAST
QUAY
WEST
PIER
SOUTH
PIER
A12
ALBERT QUAY
TANKS
(3 x 2500t)
W2
W1
Office
Fuel Loading
Points
NORTH BASE
JETTY
Min water 10m
NORTH BASE
BREAKWATER
South Base, Peterhead
Bulk Products Storage
Transit Sheds
Berth 1
Berth 2
Berth 3
Berth 4
Berth 5
Berth 6
Berth 7
Berth 8
Minimum Water Depth
6.6m
South Quay
Marina
Precision Tube Technology
Oil Tanks
Quayside Access
Key
Oil Tanks
Office Access Road
Base Open Storage
Engineering Workshop
Car Park
Office Block

Minimum Personal Protective Equipment Requirements

NATURE OF WORK	PROTECTIVE & PREVENTITIVE MEASURES REQUIRED										
	HAND	HEAD	EYE	EAR	FACE	FOOT	FALL	FLOTATION	HIGH VISIBILITY	RESPIRATORY	ADDITIONAL
Chemical Handling	1	2	1		1	1				2	Apron may be required
Cleaning, Heavy Duty	1	2	1		2	1				2	Long sleeved coveralls
Cleaning, Light Duty	1	2	1			1					
Deck Work, Anchor Handling	1	1	1	2		1		1	1		
Deck Work, General	1	1	1	2		1			1		
Deck Work, Tending Pilot	1	1	1	2		1		1	1		
Electrical Work	1	2	1	2		1			2		Insulating mat may be required
Enclosed Space Entry	1	1	2	2		1	2		2	2	Long sleeved coveralls
Food Preparation	1					1					Apron may be required
Hot Work	1	2	1	2	2	1					
Machinery Spaces, General Work	1	2	1	1		1			2		
Using Manual Tools	1	2	1	2		1					
Using Power Tools	1	2	1	2	1	1					
Welding	1	2	1	2	1	1				2	Apron and spats required
Working at Height	1	1	1	2		1	1	2	2		Tool Belt required
Working Overside	1	1	1	2		1	1	1	2		Tool Belt required

KEY

1	=	Mandatory
2	=	As determined by risk assessment, company procedures or by the nature of the work

Annex E

Operating in Environmentally Extreme Conditions

ENVIRONMENTALLY EXTREME CONDITIONS

This Appendix relates to the following conditions:

1. Air temperatures of less than -10°C.

   Please note: however, that navigation in ice is NOT included. Guidance in this matter should be sought from other sources.

2. Air temperatures greater than +35°C.
3. Safety-Critical Equipment in Environmentally Extreme Conditions
4. Operations in Extended Periods of Darkness

COLD WEATHER CONDITIONS (< -10°C)

Personnel Considerations. Working in cold weather environments has significant implications on human capabilities, and unless proper precautions are made, these can be hazardous to a person's health. In recognition of these implications on human health and performance due to working in cold climes:

1. Basic information on human performance and health hazards when working in cold conditions.
2. Guidance for design or selection of clothing.
3. Information that can be used to help generate cold weather operations safety and operating procedures.
4. Information that can be used to preserve the health of persons working in cold environ-

ments.

5. The information that follows is provided for those owners, or operators to consider in the course of ship operation.

Human response to cold exposure. The core (trunk) of the human body should remain within a small temperature range for healthy function. Excessive cooling or excessive heating will result in abnormal cardiovascular and neurological function. The skin is the organ through which a person regulates body temperature. With an average skin temperature of 33°C (91.4°F), conductive heat loss occurs at temperatures below this value, therefore, it is easy to see how cold weather performance can significantly influence normal body function. As a person cools metabolism is increased to generate more body heat – as cooling continues a person will begin to "shiver" – a visible sign that body cooling has progressed beyond a comfortable level. Increased metabolism will reduce the amount of time a person can sustain work. Safe manual materials handling tasks require the use of sense of touch, hand dexterity, strength, and coordination. Decreases in the ability to produce force, exhibit fine control over objects, and sustain muscular work loads occur in cold working environment. Work in cold environments is related to an increased risk for musculoskeletal injury. Motor function impairments of the arms and hands will occur long before cognitive or hypothermia related disabilities occur. Impaired cognitive performance will lead to poor decision-making and increased risk for accident. Persons suffering from arthritis or rheumatism will generally experience increased levels of pain during cold weather operations.

Wind chill effect. Wind chill is the perceived decrease in air temperature due to the flow of cold air over the body. Heat is lost from the human body through a variety of processes, including convection, conduction and radiation. In still conditions the air immediately next to exposed skin heats up forming an insulating boundary layer. Air movements disrupt this boundary layer, allowing new, cooler air to replace the warmer air immediately next to the skin, resulting in the apparent cooling effect. This effect is accentuated as the wind speed increases. The effects of wind chill are illustrated in *Figure E-1*. Colour coding in *Figure E-1* relates to the potential for onset of frostbite in exposed skin.

Effect of cold exposure on cognition and reasoning ability. Tasks requiring vigilance may be hampered after prolonged exposure to cold. Decision verification procedures should be implemented. Cold weather operations, coupled with other physical distracters, such as noise or motion environments, will influence the quality of perception, memory and reasoning and compound the risk of decision-making error.

Health hazards associated with cold conditions. The list of potential injuries and issues for occupational work in cold environments is lengthy. Personnel should have adequate training to enhance preparation for work in cold environments. Proper planning and precaution can deter the potential risks of cold work. *Hypothermia.* Hypothermia is a rapid, progressive mental and physical collapse due to the body's warming mechanisms failing to maintain normal body temperatures. While hypothermia is often associated with immersion in cold water, it can also occur in air when suitable cold weather protection is not employed. Conditions of extremely low dry-ambient temperature or mildly cold ambient temperatures with wind and dampness can lead to a general cooling effect on the body. If metabolic heat production is less than the gradient of heat loss to the environment hypothermia becomes an issue.

Monitoring environmental conditions. Working in cold environments requires an understanding of the interaction between ambient temperature, wind speed, relative humidity, personnel protective equipment and task being performed. In order to limit the risk during operational activities due to cold stress and further prevent local cold injuries and general freezing, specific preventative measures should be evaluated and introduced during the planning and execution of the daily work activities. Climatic metrics such as temperature, wind speed, and humidity should be regularly monitored in the locations where outside work is to be performed. Of primary importance is a regular reporting of the wind chill or equivalent temperature. Regular communications should be maintained regarding allowable time to work outside. Indoor personnel should regularly monitor outside workers so best work-to-rest / warming schedules are maintained.

Clothing and personal protective equipment. For appropriate protection / isolation against cold

				Frostbite highly likely within 30 minutes, particularly if skin is already cold.
				Frostbite will occur within 10 minutes or less, particularly if skin is already cold
				Frostbite will occur within 2 minutes or less, particularly if skin is already cold

WIND SPEED				AIR TEMPERATURE (Celsius)												
m / s	knots	BEAUFORT (Approximate)		10	5	0	-5	-10	-15	-20	-25	-30	-35	-40	-45	-50
		Force	Description													
1	2	1	Light Air	10	5	-1	-6	-12	-17	-23	-29	-34	-40	-45	-51	-56
2.5	5	2	Light Breeze	9	3	-3	-9	-15	-21	-27	-33	-39	-45	-50	-56	-62
5	10	3	Gentle Breeze	8	1	-5	-11	-17	-24	-30	-36	-42	-49	-55	-61	-67
7.5	15	4	Moderate Breeze	7	0	-6	-13	-19	-26	-32	-38	-45	-51	-58	-64	-71
10	20	5	Fresh Breeze	6	0	-7	-14	-20	-27	-34	-40	-47	-53	-60	-67	-73
12.5	25	6	Strong Breeze	6	-1	-8	-15	-21	-28	-35	-42	-48	-55	-62	-69	-75
15	30	7	Near Gale	5	-2	-8	-15	-22	-29	-36	-43	-50	-56	-63	-70	-77
17.5	35	8	Gale	5	-2	-9	-16	-23	-30	-37	-44	-51	-58	-65	-72	-78
20	40			5	-2	-9	-16	-23	-31	-38	-45	-52	-59	-66	-73	-80
22.5	45	9	Severe Gale	4	-3	-10	-17	-24	-31	-38	-45	-53	-60	-67	-74	-81

climate conditions, adequate clothing should be selected and used on board during cold periods. Such optimal clothing should be able to mitigate water and humidity during work and at the same time insulate sufficiently to maintain thermal comfort during rest. The insulating effect of the clothing is influenced by different factors including temperature, wind and humidity. Specific guidance is to be provided covering:

- Hand Protection.
- Head and Eye Protection.
- Foot Protection.

Nutritional considerations in cold climates. The added weight of protective clothing and the limitations in mobility created by protective equipment will increase the mobility demands of the operator, thus increasing the metabolic needs for a given task.

Workstation design and operational considerations. An analysis of outdoor work situations should be performed early in design / layout development, and should be updated when design changes are made that will influence personnel's exposure to cold stress. Outdoor operations analyses (an examination of the tasks to be carried out in cold conditions) should be carried out for open work areas and semi-open work areas. The objective of these analyses is to identify and remedy task performance issues due to overall exposure to temperature, wind, icing and precipitation, including investigation of the weather protection necessary to comply with exposure limits.

Safety systems. Cold environments present many significant challenges to the design and use of emergency, evacuation, and rescue devices. Much of the hardware devised for such use is designed for more temperate climates. Fire mains can freeze. Materials (such as used in life vests) become brittle. Working devices (such as sheaves, blocks, and davits) can freeze in place – refusing to move.

Fire fighting equipment. Significant risks are associated with fire fighting equipment, the most significant being the potential freezing of fluids in lines, thereby depriving crew of the use of the fire fighting systems. Specific risks include:

- Freezing of fire water hoses, piping, nozzles, etc.
- Portable fire extinguisher storage may be obstructed or frozen.
- Fire dampers may freeze in the stowage position (generally closed in temperate climates).

Appliance types include lifeboats, life rafts, rescue boats, launching stations, ice gangways, immersion suits, alarms, escape routes, and access routes.

Hull construction, arrangement, and equipment. Specific features which should be included when planning operations in cold conditions include:

- *Ballast Tanks.* Means must be provided to prevent freezing of the ballast water in tanks and vents.
- *Superstructure and deck houses.* External access to the navigation bridge windows is to be provided to facilitate ease of cleaning. Alternating navigation bridge windows are required to be heated.
- *Personnel required to perform external duties* such as being a lookout when underway, security at the gangway when in port, or being on deck during loading operations are to be provided with a safe haven.

Ice loads on decks. In particular, one of the potentially significant consequences for any ship in transit through cold weather waters is the concentration of ice on deck.

Seawater supplies. During navigation and at port in ice-covered waters, attention must be paid to sea water supplies for essential operational systems and safety systems. Sea water supplies are needed for the ballast system, the cooling water system serving propulsion machinery, main and emergency fire pumps supplying the fire and wash deck system and the water spray system.

Protection of deck machinery, systems and equipment. Generally, deck machinery and systems are not prepared for freezing temperatures. Essential equipment and systems must be available at all times and in any temperature conditions. The lubricating oil and hydraulic oil used in rotating machines exposed to the weather must be suitable for low temperatures.

WARM AND HOT CONDITIONS (> 35°C)

Ultra violet protection (personnel). Personnel should be made aware of the risks associated with excessive exposure to ultra violet radiation. It should be noted that the risk of over-exposure to ultra-violet radiation is not limited to warm / hot conditions, and may also occur in middle or high latitudes. Operations should be planned or equipment provided so that the risk of personnel being exposed to excessive or extended radiation is minimised.

Ultra violet protection (equipment). Certain items of equipment, including some plastics and ropes manufactured using artificial fibres, will quickly degrade when exposed to intense ultra violet radiation, leading to failure in use which may result in a dangerous situation developing. Arrangements should therefore be made to protect such equipment form exposure to radiation of this type. It may be inevitable that equipment cannot be protected whilst in use, but every effort should be made to provide adequate protection when not actually deployed.

Precautions against infection. Operations in certain tropical or equatorial parts of the globe may result in personnel being exposed to the risk of contracting infections or diseases against which their natural immune system will provide little or no defence. Personnel should therefore be made aware of such risks and the measures to be taken to minimise them. Where appropriate, barrier arrangements, insect repellents and prophylactic medicines should be provided. Personnel are to be instructed in their use as required.

Precautions against dehydration. Dehydration, with its associated risks, may be experienced by personnel engaged in strenuous activities which result in increased perspiration. Whilst this may occur in temperate climates this risk increases in tropical and equatorial conditions, particularly since even minor levels of activity may give rise to excessive perspiration. Personnel should therefore be made aware of the risks associated with dehydration, and sufficient drinking water should always be made readily available. As described in standard medical reference sources urine colour is an easy way to monitor an individual's hydration status. These should be consulted for further information regarding this matter.

SAFETY CRITICAL EQUIPMENT IN ENVIRONMENTALLY EXTREME CONDITIONS

Marine equipment, including that of a safety critical nature, is normally designed and manufactured to operate within a temperature range from -10°C to +35°C. There there is any likelihood that vessels will be required to operate in conditions outside this temperature range it should be ensured that all elements of safety critical systems are designed and manufactured accordingly, or are adequately protected to ensure their continuing operability. A programme of regular inspection and testing of safety critical systems when operating outside the normal temperature range should also be implemented to ensure that such systems remain available if required.

OPERATIONS IN EXTENDED PERIODS OF DARKNESS

Operations in higher latitudes (north and south) in the winter months will be undertaken in circumstances where the period of natural daylight is restricted or absent altogether and will involve extensive use of artificial illumination. In such circumstances due recognition should be taken of the risk that personnel will experience depression or other adverse effects due to seasonal affective disorder (SAD). Clinical advice should be sought to identify the appropriate precautions to be taken to minimise this risk.

Safety Zone Entry Check Lists

ALL VESSELS, ARRIVAL AT OFFSHORE FACILITY

Vessel	
Facility	
Date and Time	

	All Vessels	**Status**		**Comments**
		Yes	**No**	
1	Environmental conditions acceptable for a safe operation (including wind, sea, swell, visibility and current)			
2	Limitations due to sea / weather condition			
3	Safe approach / exit routes identified Stand Off location identified			
4	Confirm whether any simultaneous operations anticipated whilst vessel is within safety zone			
5	Confirm whether any prohibited zones at facility			
6	6 Bridge and Engine room manned in accordance with GOMO			
7	Communication established VHF Channel(s): UHF Channel(s):			
8	No hot work / smoking on deck within safety zone			
9	Auto Pilot off			
10	All manoeuvring and steering gear systems tested including changeover between control positions and manoeuvring modes			
11	Emergency manoeuvring system confirmed to be operational			
12	Operating location confirmed with facility			
13	Status of overside discharges confirmed with facility			
14	Vessel to be manoeuvred to set-up position before changing mode (1.5 ~ 2.5 ship's lengths depending on whether in drift on or drift off situation)			
15	Vessel operational capability reviewed / confirmed (to include power, thrust, location, heading, etc.)			
16	Risk assessment for alongside operations reviewed / confirmed (If working on weather side, complete additional RA)			
17	Facility to confirm readiness for vessel arrival and operation (including no overboard discharge)			
18	Manoeuvring mode during the operation to be agreed (If DP mode vessel specific DP checklist to be completed)			
19	On-going and / or planned activities within safety zone confirmed between facility and any other vessels			

FURTHER CHECKLISTS TO BE COMPLED AS APPROPRIATE

PERMISSION RECEIVED TO ENTER SAFETY ZONE			
DATE		**TIME**	
FROM		**FUNCTION**	
NAME		**NAME**	
SIGNATURE		**SIGNATURE**	
POSITION / RANK		**POSITION / RANK**	

VESSELS ENGAGED IN LOGISTICS SUPPORT

IS THIS CHECKLIST RELEVANT: YES / NO

(Delete as appropriate   if not relevant, cross out the checklist)

Vessels Engaged in Logistics Support	Status		Comments
	Yes	No	
1 Proposed operations confirmed with facility. Discharge and Back-Load (Cargo, bulks, fluids, etc)			
2 Anticipated duration of operations confirmed			
3 Confirm discharge / back-load sequence with facility 1. Can stow be broken safely ? (Adequate escape routes to safe havens, etc.) 2. Any priority lifts ? (Must not require "cherry picking" of stow) 3. Sufficient space for back-load except at last call ? (See Notes below)			
4 Confirm any other activities which may occur whilst vessel is alongside and connected to facility (Particularly any operations involving crane driver or deck crew)			
5 Confirm availability of facility personnel, equipment (Particularly for any operations involving hoses)			
6 Confirm whether any changes of working face will be required (If so, move from one to the next to be planned accordingly)			
7 Communication established VHF Channel(s): UHF Channel(s):			
8 No hot work / smoking on deck within safety zone			
9 Auto Pilot off			
10 All manoeuvring and steering gear systems tested including changeover between control positions and manoeuvring modes			
11 Emergency manoeuvring system confirmed to be operational			
12 Operating location confirmed with facility			
13 Status of overside discharges confirmed with facility			
14 Vessel to be manoeuvred to set-up position before changing mode (1.5 ~ 2.5 ship's lengths depending on whether in drift on or drift off situation)			
15 Vessel operational capability reviewed / confirmed (to include power, thrust, location, heading, etc.)			
16 Risk assessment for alongside operations reviewed / confirmed (If working on weather side, complete additional RA)			
17 Facility to confirm readiness for vessel arrival and operation (including no overboard discharge)			
18 Manoeuvring mode during the operation to be agreed (If DP mode vessel specific DP checklist to be completed)			
19 On-going and / or planned activities within safety zone confirmed between facility and any other vessels			

FURTHER CHECKLISTS TO BE COMPLED AS APPROPRIATE

PERMISSION RECEIVED TO ENTER SAFETY ZONE			
DATE		TIME	
FROM		FUNCTION	
NAME		NAME	
SIGNATURE		SIGNATURE	
POSITION / RANK		POSITION / RANK	

ALL VESSELS, DEPARTURE FROM OFFSHORE FACILITY

	Vessels Engaged in Logistics Support	Status		Comments
		Yes	No	
1	Vessel to be manoeuvred well clear of Facility before changing Mode (1.5 ~ 2.5 ship's lengths depending on whether in drift on or drift off situation)			
2	All controls set to neutral position before changing mode			
3	Where practical, vessel to depart down weather or down current from facility			

CHECK LIST FOR OFFSHORE FACILITIES

ALL VESSELS ENTERING SAFETY ZONE

Vessel	
Facility	
Date and Time	

	All Vessels	Status		Comments
		Yes	No	
1	Confirm anticipated working locations with vessel			
2	Confirm status of facility, where relevant (Heading, movement, thruster use, etc.)			
3	Communication established: VHF Channel(s): UHF Channel(s):			
4	Confirm that vessel is aware of any prohibited zones around the facility			
5	Confirm whether any simultaneous operations anticipated whilst vessel is within safety zone (nature and duration of any such operations to be advised to vessel)			
6	Status of overside discharges to be confirmed and advised to vessel			
7	Vessel operational capability confirmed (Vessel to advise any concerns and / or operational limits)			
8	Facility to confirm readiness for vessel arrival and operation (Including overboard discharges stopped where practical)			
9	Vessel to advise proposed station keeping arrangements during the operation (if DP mode confirm proposed operational mode)			
10	Confirm names of any other vessel attending the facility 1. SBV: 2. Vessel 1: 3. Vessel 2:			

FURTHER CHECKLISTS TO BE COMPLED AS APPROPRIATE

PERMISSION GIVEN TO ENTER SAFETY ZONE			
DATE		TIME	
NAME		NAME	
SIGNATURE		SIGNATURE	
POSITION		POSITION	

VESSELS ENTERING THE SAFETY ZONE FOR LOGISTICS SUPPORT

IS THIS CHECKLIST RELEVANT: YES / NO

(Delete as appropriate – if not relevant, cross out the checklist)

	Vessels Engaged in Logistics Support	Status		Comments
		Yes	No	
1	Proposed operations confirmed with vessel Discharge and Back-Load (Cargo, bulks, fluids, etc)			
2	Anticipated duration of operations confirmed			
3	Vessel to be informed of any anticipated delays during operations.			
4	Confirm discharge / back-load sequence with vessel 1. Are there any priority lifts? (Must not require "cherry picking" of stow) 2. Has vessel sufficient space for back-load?			
5	Confirm any other activities which may occur whilst vessel is alongside and connected to facility (particularly any operations involving crane driver or deck crew)			
6	Confirm availability of facility personnel, equipment (particularly for any operations involving hoses)			
7	Confirm whether any changes of working face will be required			
8	Confirm whether any unusual lifts will be involved 1. Any Main Block Lifts? 2. Any Vulnerable / Sensitive Lifts? 3. Any Lifts involving use of Tag Lines? 4. Any other Unusual Lifts? (Including long objects, or not pre-slung lifts, etc.) 5. Has crane driver appropriate competency and experience? (previous experience of lifts of this nature)			
9	Confirm readiness to commence bulk transfer operations 1. Is Hose buoyancy adequate ? 2. Has Valve configuration been correctly set ?			
10	Confirm readiness to commence liquid transfer operations 1. Is Hose buoyancy adequate ? 2. Has Valve configuration been correctly set ? 3. If required, is illumination adequate ?			
11	Confirm whether vessel will be required to receive any back-load bulk cargoes. (If so, confirm recent analysis report will be available prior to cargo being delivered to vessel)			

Typical Deck Cargo Plan

DECK CARGO PLAN

VESSEL:	DATE:
FROM:	VOYAGE NUMBER:
TO:	

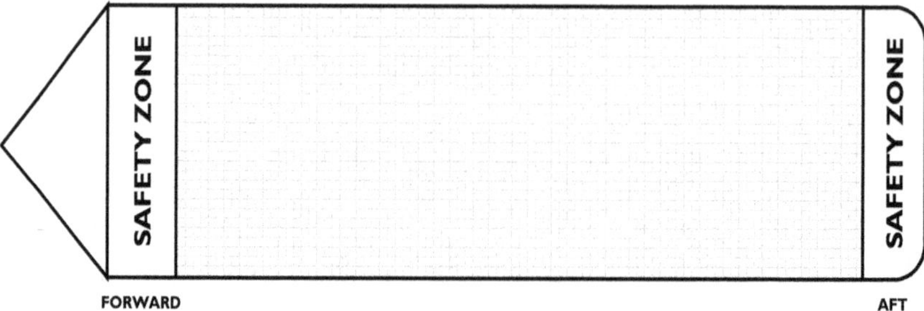

Annex H

Carriage of Tubular Cargo

GENERAL

The purpose of this document is to describe the recommended practice for safe transportation and handling of tubular cargo on offshore service vessels. Important interface issues in relation to bases and installations are included. According to governing regulations, it is the responsibility of the master to make sure that the cargo is properly secured before the departure. This document does not in any way or manner exempt the master from this responsibility but is intended to serve as the recommended practice for handling of tubular cargo on vessels in connection with cargo handling at bases and offshore as well as during transport.

Definitions, Images and References

Tubular cargo is defined as any type of round objects which are shipped not in separate cargo carriers but using slings to bundle one or more such objects together in a bundle.

Pup joints are short casing or tubing joints used as "space out" for connecting pipeline sections of a pre-determined length.

Centralisers are devices fitted to the outside of the casing / liner to align it to the centre of the bore hole during cementing.

CARGO REQUIREMENTS

The slinging shall be in accordance with national requirements and branch standards, and proper secured with wire clamps or similar ex. Welcro bands. Units shorter than six metres should be transported in a cargo units. Slinging of tubular cargo must ensure the bundles remain stable. Tubular cargo should preferably be bundled in odd numbers where practicable. As regards 9 5 / 8" – 13 3 / 8" casings fitted with a centraliser, consider having only one tubular in each bundle as it may be difficult to split them on the pipe deck. The slinger must take into consideration the WLL of the slings and the weight of each tubular when slinging the bundles. Certified lifting points fitted on the tubular cargo shall be used during loading of large and heavy dimensions if they cannot be strapped in a prudent manner or handled in certified cargo carriers. Inspect for loose and or damaged protectors during all phases before lifting the cargo.

REPARATIONS BEFORE LOADING AT BASE

The vessel must be informed of the tubular cargo well before loading including the dimensions, weight, length, and quantity. Dedicate the most suitable deck area based on destination, which crane will be used and weather reports. And, if relevant, how many layers may be loaded. Hull loads and reduced stability in case tubular cargo become filled with water must be taken into consideration upon assignment of area.

- Position hawsers: three hawsers are recommended across the deck for each joint, one approx. in the middle, and one 1-2 metres from each of the ends.
- Position chains: two chains are recommended below the first layer for each joint, about 1 / 3 and 1 / 4 of the distance from each end. It may not be necessary to use the chain during the loading. But if the offloading operation must be interrupted before all cargo has been unloaded, the chains may be used to secure the remaining cargo
- Prepare pipe supports. The vessels will normally have pipe supports approx. 1 / 3 in from the cargo rail on each side.
- Prepare Automatic Sea Fastening Arrangement if necessary on vessels equipped with this.
- Pay special attention during loading on steel decks on anchor handling vessels. The vessel crew must position enough friction material (hawsers) before the loading starts.
- Chains must also be used.

A sufficiently safe zone must be established fore and aft of the dedicated cargo area. The area must be a minimum of one metre.

LOADING AT BASE

A representative from the vessel, preferably an officer responsible for loading, must monitor and supervise during the loading operation. It is important to ensure bundles are stowed as close together as possible to avoid the risk of shifting cargo during the voyage. When loading large dimensions with one tubular in each bundle, evaluate whether to fit wedges below each tubular joint to avoid the risk of shifting cargo during transportation or offloading. If wedges are used, these should be nailed to a wood deck if possible to reduce the risk of shifting. Large dimensions must never be loaded on top of smaller dimensions. When stacking cargo, take into consideration the strength of the deck, as well as the working height for seamen. Two metres is normally the maximum stacking height. Vessels must always be loaded in a manner that make it possible easy securing of remaining cargo on board in case of interrupted offloading offshore. If possible, tanks and other frame / skid-type cargo units shall not be positioned just fore or aft of tubular cargo due to the risk of snagging. Slings on bundles must be extended and laid across the tubular cargo to avoid becoming wedged between the bundles. Determine the appropriate placement in relation to openings and escape routes in cargo rails, etc. Cargo units shall not be used as the only barrier to secure tubular cargo on vessel decks.

TRANSPORT

The risk of shifting cargo is normally highest during the voyage / sailing to from an installation. In the event of marginal weather conditions, the risk of shifting tubular cargo must be taken into account when selecting the time of departure, route and speed.

PREPARATIONS FOR OFFSHORE LOADING / OFFLOADING OPERATIONS

Conduct an internal Pre-Job Talk on the vessel to assess / clarify the following as a minimum:

- Communications.
- Positioning of the vessel.
- Distribution of work / roles between the seamen on deck when two pendants / hooks are used.
- Operation-specific issues such as the weather, type of tubular cargo, location, any securing arrangements.

Conduct a Pre-Job Talk between the vessel and the crane operator to clarify the following as a minimum:

- Communications.
- How many bundles for each lift (recommended two bundles).
- Any use of tag lines during offloading to the installation.
- Positioning of the vessel as regards vessel movements, reach and line of sight from the crane.
- Operation-specific issues, including risk of snagging.

OFFLOADING AT INSTALLATION

Pay special attention during removal of any lashings used during the voyage out to the field. It is important to use correct footwear (protective footwear covering the ankles) if anyone has to walk on top of tubular cargo. Focus on correct dogging. Recommended two eyes in each hook depending on lifting equipment. The deck crew, hook and cargo on the vessel deck must be within line of sight of the crane operator. Good radio discipline is important – "talk where the hook is". Avoid the use of tag lines if possible. If tag lines must be used, fasten and prepare these before dogging of the individual lifts. The risk of snagging on the vessel deck and cargo rails, as well as in potential blind zones, must be taken into account during positioning of the vessel.

LOADING TO VESSEL AT INSTALLATION

In addition to the issues addressed under section 7 of the Guidelines; Offloading at installation, the following issues are important during loading onto vessels at the installation:

- Vessels must be informed of the type, quantity and weight to be returned well before loading starts.
- The vessel crew must prepare the necessary hawsers as well as chains and pipe supports.
- All tubular cargo to be returned to a vessel should be washed first to avoid slippery tubular cargo on the vessel deck.
- It is important to use correct footwear (protective footwear covering the ankles) if anyone has to walk on top of tubular cargo.
- Tubular cargo shorter than six metres should be shipped in baskets.
- If possible, avoid tubular cargo where the crew of the vessel must unhook / hook lifting yokes.
- Tag lines should not be used during loading of return cargo onto vessels.
- The crew of the vessel must never touch lifts of tubular cargo or walk underneath such lifts before the lift has been landed properly.
- Slings on bundles must be extended and laid across the tubular cargo to avoid becoming wedged between the bundles.
- During loading of return cargo, pay special attention to rolling cargo. In connection with large dimensions and if the vessel is rolling, any vessel without ASFA or equivalent must use wedges to secure large dimension cargo before unhooking it. It may be useful to have the

vessel list somewhat towards the side where the first lifts will be landed.

INTERRUPTED OFFLOADING / LOADING AT INSTALLATION

In the event of interrupted offloading or loading at the installation, the vessel must be able to and be given enough time to secure the remaining cargo in a proper manner.

OFFLOADING AT BASE

The deck crew must be careful during removal of sea lashings upon arrival at the base. If other cargo is placed adjacent to tubular cargo upon arrival at the base, pay special attention during offloading of this cargo.

LOADING OF TUBULAR CARGO ONTO PIPELAYING VESSELS

Loading of tubular cargo for pipeline installation projects are normally handled by the pipelaying contractor chartering and employing the vessel, and not by the technical developer. Lifting beams are normally used during offloading of this type of tubular cargo, and they are lifted by inserting each end of the tubular cargo to be lifted into the lifting equipment. During loading of large quantities of tubular cargo onto pipelaying vessels, take into consideration that the seamen need a safe workplace as well as the maximum total cargo that the vessel can hold. In the event of large heights, start loading from the middle to avoid work towards the outer perimeter of the cargo deck (risk of falling overboard?) (3) Hull loads and reduced stability resulting from weight of tubular cargo, including water inside and between them, must be included in the stability calculations. (4) The *Monsvik Method* for loading of tubular cargo prevents very large open spaces between the pipeline bays. The distance down to the deck with 4 or 5 tubular cargo stacked on top of each other may be several metres. A fall may prove fatal.

Figure H-1 Illustration of the Monsvik Method for loading tubular cargo

Fig.H-2 Marine Risers

Fig.H-3 Conductors (dimensions range from 26" to 32")

Fig.H-4 Casing (dimensions range from 7" to 26")

Fig.H-5 Drill Pipe

Fig.H-6 Slip Joint

Fig.H-7 Drill Collars

Fig.H-8 Tubing (dimensions range from 2 7 / 8" to 7")

Fig.H-9 Tubular cargo for Pipelaying Vessels (transport pipeline)

Fig.H-10 ASFA (Automatic Sea Fastening Arrangement)

Make Up and Use of Tag Lines

In certain circumstances light, soft lines may be used to assist in the handling of long and or fragile items of cargo. These are often referred to as "tag" lines. It must be recognised that whilst such aids may assist operations their use does introduce some additional risks, as described below.

RISKS

Additional risks associated with the use of tag lines include potential injuries from dropped objects as a result of personnel handling cargo having to work in closer proximity to suspended loads than would normally be the case; potential injuries resulting from personnel handling cargo being dragged across the handling area through a heavy load rotating in an uncontrolled manner and the tag line being fouled in limbs or clothing; and potential injuries resulting from tag lines being secured to adjacent fixed structures parting and whipping back as a result of a heavy load rotating in an uncontrolled manner.

MITIGATION OF RISKS

Make-up of lines. Tag lines must be made up from single, continuous lengths of rope. Apart from the knot attaching the line to the cargo, there must be no other joints or knots in the line. Tag lines must be of sufficient length to allow personnel handling cargo to work in a safe position well clear of the immediate vicinity of the load. It this regard it is recommended that the length of the line should be not less than 1.5 times the maximum height above the handling area at which the arrangements will be used.

In use. Whilst in use, precautions should be observed as follows:

1. Tag lines are an aid to positioning the load when landing, and as such must only be used when weather conditions would permit the lifting of the item without the use of such arrangements. It must not be assumed that in conditions more severe than this the use of tag lines will allow the operation to be completed safely.
2. At all times personnel handling tag lines must work at a horizontal distance from the load equivalent to its height above the handling area, maintaining an angle between the line and the horizontal of not more than 45 degrees.
3. All sections of the line, including slack must be kept in front of the body, between the handler and the load.
4. Where two or more persons are handling the same line, ALL must work on the same side of the line. Any slack must be kept in front of the group.
5. Tag lines must be held in such a manner that they can be quickly and totally released. They must not be looped around wrists, or other parts of the body.
6. Particular care must be taken when using tag lines whilst wearing gloves, to ensure that the line does not foul the glove.
7. Tag lines must not be secured or attached in any manner to adjacent structures or equipment. This includes the practice of making a "round turn" on stanchions or similar structures and surging the line to control the load.
8. Where pre-installed lines are used consideration should be given to providing personnel with boat hooks or similar equipment to retrieve the lines without having to approach the dangerous area in the vicinity of the suspended load.
9. An example of such circumstances would be when lines are attached to a load on the deck of a vessel, the load being then transferred to an offshore installation.

Wet Bulk Cargo Transfer Check Lists

WET BULK TRANSFER CHECK LIST

PRE-START CHECK LIST: PORT		
1	Type and quantity of product(s) to be transferred, confirmed and MSDS available	
2	Allocate tanks to product	
3	Confirm transfer rate and max. allowable rate per product	
4	Topping off procedure agreed	
5	Emergency stop procedure agreed	
6	Hose(s) confirmed as fit for purpose and of sufficient length	
7	Hose(s) connected to correct coupling(s)	
8	Vessel springs tensioned to limit ranging	
9	Communications procedure established for transfer, including agreement on central control point, i.e., the bridge	
10	Appropriate pollution prevention equipment deployed as SOPEP	
11	Scuppers plugged if hydrocarbons to be transferred	
12	All Hot Work Permits withdrawn if hydrocarbons to be transferred.	
13	Self-sealing couplings to be used if fuel to be transferred	
14	Lines set ready for cargo transfer	
15	Tank monitoring system proven	
16	Watch established on manifold with suitable communications in place	

PRE-START CHECK LIST: OFFSHORE

1	Type and quantity of product(s) to be transferred, confirmed and MSDS available		
2	Order of discharge confirmed, if more than one		
3	Confirm transfer rate and max. allowable rate per product		
4	Emergency stop procedure agreed		
5	Tank changeover / topping off procedure agreed		
6	Confirm notice required to stop cargo		
7	Confirm whether vessel or installation stop		
8	Slings and lifting arrangement satisfactory		
9	Hose(s) visually inspected and found suitable		
10	Hose(s) connected to correct coupling(s)		
11	Communications procedure established and agreed for transfer		
12	Appropriate pollution prevention equipment deployed as per SOPEP		
13	Underdeck lighting adequate for task in hand		
14	One person appointed to sight hose(s) and advise Master of position		
15	Lines set ready for transfer		
16	Crane Operator and both installation and vessel deck crews close at hand		

TRANSFER CHECK LIST: PORT

1	All communications to be routed via control point which should be vessel bridge	
2	Start transfer slowly until cargo confirmed as entering correct tank(s)	
3	Volume checks conducted at regular intervals with receiver / provider	
4	All personnel involved in transfer in regular contact	
5	Adequate warning given of tank changeover	
6	Rate reduced for topping off	

TRANSFER CHECK LIST: OFFSHORE

1	Start transfer slowly until cargo confirmed as entering correct tank(s)	
2	If fuel to be transferred, line checked for leaks at start up	
3	Volume checks conducted at regular intervals with receiver	
4	Cargo Officer can see bulk hose(s) throughout	
5	Adequate warning given of tank changeover etc.	

Dry Bulk Cargo Transfer Check Lists

PRE-START CHECK LIST: LOADING

1	No residue remaining from previous cargo and tank (s) dry	
2	Tank air distribution slides are in good condition	
3	Tank access seals are in good condition	
4	Type and quantity of product(s) to be loaded confirmed and MSDS available	
5	Tank(s) allocated to product	
6	Order of loading confirmed, if more than one product to be loaded	
7	Proper vent line connected to vessel	
8	Confirm loading rate and max. allowable rate per product	
9	Emergency stop procedure, agreed	
10	Notice required to stop, agreed	
11	Confirm whether cargo will be stopped by vessel or provider	
12	Confirm tank(s) and lines are vented to atmospheric pressure	
13	Confirm Lines set for cargo	
14	Hose(s) connected to correct coupling(s)	
15	Hose(s) inspected and fit for purpose	
16	Moorings tensioned sufficiently, particularly springs, to limit ranging	
17	Communications procedure established for transfer, including agreement on central control point, i.e., the bridge	
18	Watch established on manifold with suitable communications in place	

PRE-START CHECK LIST: DISCHARGING

1	Vessel settled in position and ready to receive hose(s)	
2	Type and quantity of product(s) to be transferred confirmed and MSDS available	
3	Appropriate tankage on vessel lined up and ready for discharge	
4	Confirm transfer rate and max. allowable per product	
5	Emergency stop procedure, agreed	
6	Notice required to stop, agreed	
7	Confirm whether cargo will be stopped by vessel or receiver	
8	Hose Lifting arrangement satisfactory	
9	Hose(s) visually inspected and found fit for purpose	
10	System de-pressurised, ready for hose(s)	
11	Hose(s) connected to correct coupling(s)	
12	Communications procedure established and agreed for transfer	
13	Underdeck lighting adequate task in hand	
14	Vent position(s) identified	
15	Cargo Officer appointed to watch hose(s) relative to vessel's stern	
16	Crane Operator and both installation and vessel deck crews close at hand	

LOADING CHECK LIST

1	All communications to be routed via control point which should be vessel bridge	
2	Good vent obtained on start up	
3	Bulk hose(s) and vent checked throughout operation for blockages	
4	Contact with loading personnel maintained throughout	
5	Lines cleared back to vessel	
6	System de-pressurised on completion, before disconnection	

DISCHARGING CHECK LIST

1	Good vent obtained from receiver before commencing discharge of cargo	
2	Good watch maintained on hose(s) in case of blockage	
3	Contact with receiver's personnel maintained throughout	
4	Lines blown clear to receiver on completion of cargo	
5	System de-pressurised before hose disconnection	
6	Blank cap(s) fitted to hose end(s) before passing back to receiver	

Tank Cleaning Check Lists

TANK CLEANING CHECK LIST

Check List Nos			Vessel Name			Vessel Permit Nos		
Reason for Entry								
Tank Nos								
Confined Space Contents								

SAFETY CHECKS

1	Has enclosed space been thoroughly:	Yes	N/A	7	Hazards	Yes	N/A
1.1	Depressurised			7.1	Noise		
1.2	Ventilated (by natural / mechanical means)			7.2	Toxic		
1.3	Drained			7.3	Chemical		
1.4	Isolated by - Blanking			7.4	Corrosive		
	- Disconnecting			7.5	Explosive		
	- Valves			7.6	Flammable		
1.5	Steamed			7.7	Electrical		
1.6	Water Flushed			7.8	Static Electricity		
1.7	Inert Gas Purged			7.9	Fall from Height		
1.8	Tank Appliances Electrically Isolated and Locked			7.10	Overhead Hazards		
1.9	Opened tank hatches guarded			7.11	Potential Dropped Objects		
2	Tank Prime Mover has been:	Yes	N/A	7.12	Entrapment		
2.1	Electrically isolated and locked (all stations, wheelhouse engine room, etc.)			7.13	High Pressure Jetting		
2.2	Mechanical motive power isolated and locked (all stations, wheelhouse engine room, etc.)			7.14	Suction		
3	Vessel Machinery – Main Engines, Shafts Generators etc.	Yes	N/A	7.15	Trip Hazards (specify)		
3.1	Agreed isolation of vessel machinery (all stations, wheelhouse engine room, etc.)			7.16	Hot Surfaces		
3.2	Agreed change of status, start-up procedure			7.17	Other		
4	Other Considerations:	Yes	N/A	8	Protective Equipment		
4.1	Material Safety Data Sheets available			8.1	Eye Protection (Specify)		
4.2	Annex 10-E-2 Sheets			8.2	Face Shields		
4.3	Suitable Access / Egress provided			9.3	Respirator		
4.4	Standby Personnel detailed			8.4	PVC Gloves		
4.5	Lifeline / Safety Harnesses / Rescue Hoist			8.5	Safety Boots		
4.6	Breathing Apparatus			8.6	High Pressure Jetting Boots		
4.7	Means of communications tested OK			8.7	Wet Suit		
4.8	Area free of flammable materials			8.8	Full Chemical Protective Clothing		
4.9	Area free of Ignition sources			8.9	Breathing Apparatus		
4.10	Work time / fatigue			8.10	Head Protection		
4.11	Clear working area			8.11	Ear Protection		
4.12	Illuminations			8.12	Other		
4.13	Visibility of Hoses			9	Emergency Procedures		
4.14	Other work that could cause hazards			9.1	Muster Points Identified		
5	Toolbox Talk:	Yes	N/A	9.2	Escape Routes Identified		
5.1	TBT conducted with ALL applicable personnel			9.3	Alarms Understood		
5.2	Special Training / Briefing required			9.4	Location of Firefighting and First Aid Equipment		
5.3	Other			9.5	Contact Numbers		
6	Plant Required				Emergency Services		

6.1	Compressor		6.5	Safety Barriers / Signs	Vessel Bridge
6.2	Pressure Washers		6.6	Lighting	Base Operator
6.3	Vacuum Tankers		6.7	Air Driven Pumps	Tank Cleaning Contractor
6.4	Jetting Lance Baffles		6.8	Others (Specify)	

10	Other Requirements / Limitations:

11	Monitoring	Yes	No			Yes	No
11.1	Ongoing Gas Monitoring Required			11.3	Competent Analyst(s) Required		
11.2	Frequency of Ongoing Monitoring	30 mins	1 hour		2 hours	Other - Specify	

DECLARATION

I have personally checked the above conditions and consider it safe to enter provided that the conditions laid down are adhered to:

Tank Cleaning Contractor	Signed	Print Name	Date
Client / Vessel Master (or Designate)	Signed	Print Name	Date

Annex M

Tank Cleaning Standards

Tank inspections should confirm that the tanks have been cleaned to the following standards as required.

Brine standards. Cargo lines and pumps are flushed through with clean water and lines drained. Tank bottoms and internal structure (stringers, frames, etc.) are clear of mud solids, semi solids and all evidence of previous cargo. The tank may require cleaning with detergent to achieve the highest standard of cleanliness possible. All traces of water and detergent removed from tank.

Water based mud standard. Cargo lines and pumps are flushed through with clean water and lines drained. Tank bottoms and internal structure (stringers, frames, etc.) are clear of mud solids, semi solids and all evidence of previous cargo. The tank may require cleaning with detergent to achieve the highest standard of cleanliness possible. All traces of water and detergent removed from tank.

Oil based mud standard. Tank bottoms and internal structure (stringers, frames, etc.) are clear of mud solids and semi-solids. Cargo lines are flushed through with clean water and lines drained. Pump suctions are checked and clean. Tank must be empty and clear of all water / mud mixtures.

Pump out standard. Pump out residues from tank and wipe tank floor using rubber mops or equivalent. Check suction pipes to ensure they are clear. No requirement for washing.

Dry bulk tanks. Tanks to be brushed down and residues removed by vacuum tanker, eductor system or equivalent. Slides to be checked for dryness and condition and 'elephant foot' suction checked to be clear.

Annex N

Bulk Hose Best Practice

INTRODUCTION

Background. Integra is an initiative that was established by Sparrows and Sigma 3 in 2006 to deliver best practices in crane and deck operations in the offshore industry. It was quickly realised during their offshore visits that there was a need to produce guidance to manage bulk hose systems safely and at the same time create a common practice throughout the industry. The key elements required in coaching personnel to recognise and eliminate hazardous risks to themselves and others are contained within management systems. These systems eliminate damage to plant and equipment, providing safer operations and stricter controls of environmental issues. We recognised minimum standards of controls and guidance are in place to manage and maintain bulk hose systems, including hose hang-off points on installation structures. We also acknowledged the difficulties in being prescriptive due to the differences in installation layouts and working practices. To make positive changes in our operations we have collated information from the workforce on how to manage work involving bulk hose systems. The following guidelines indicate best practices which will reduce the number of hose failure incidents in the industry and the resulting exposure to the environment.

Environmental issues. With the evidence available it was identified that 21% of spills to the sea were hose related incidents. The most common bulk hose failures are due to abrasion to the outer cover of the hoses rubbing on the installation structures, resulting in leakage from the hose string. The wear on the hose is accelerated when the hose radius exceeds the recommended minimum bend radius criteria causing premature failure. Both examples can cause the hose to leak into the sea if not controlled by a robust hose management system. All environmentally sensitive products should have suitable hose connections or similar self sealing connection on the hose end.

GENERAL REQUIREMENTS

The following recommendations apply to any hose which carries products, including products that are harmful to the environment if containment is lost. It is recommended by hose manufacturers, based on information taken from previous incidents on installations that a bulk hose should be changed out approximately every two years due to internal fatigue to the hose layers. When the hose is not in use an end cap, commonly known as a blank, should be used on the connection that marries the hose to the vessel manifold, and where possible protect the hose ends with a waterproof cover preventing contamination, corrosion or damage to the hose connection. When hose strings are suspended from the installation, they should be suspended well clear of the sea and restrained to the installation minimising movement and abrasion to the hoses' outer cover, preventing the waves from twisting the hoses. Where the hose may contact any part of the installation structures all contact points on the hose should be covered with a form of protection. Floatation collars can be used or alternatively, sections of redundant hose can be fitted to the structures at impact abrasion points. Floatation collars should be used either side of the hose couplings to prevent the coupling damaging an adjacent hose in the fingers. To prevent excessive load on a suspended hose string, the hose should be drained back to the vessel or installation once offloading is completed. Hoses should be suspended from sound structures or certified lifting/hang-off points on the installation to prevent kinking in the hose string. If required the LOLER Competent Person or equivalent, or structural engineer, should be consulted for guidance. Hoses should never be suspended or supported by wire slings as they may cut into the hose and damage the hose structure.

The LOLER Competent Person should be consulted for selection of correctly certified and appropriate slings. When replacing a length of hose in a string, the string must be brought in board and barriers erected round the hose indicating no unauthorised entry during the replacement of the hose and/or floatation collars. Once a hammer lug union is installed and tightened it should be marked across both faces with paint or similar permanent marker to monitor the fitting is continually taut and fit for purpose. Care should be taken when using cutting tools to remove packaging from a new hose. It is imperative that no damage comes to the hose section during unpacking. Prior to commencing any offloading operations the hose string should be visually inspected for damage using the list below as a minimum check:

- Leaks at the hose fitting or in the hose make up.
- Damaged, cut or abraded covers.
- Exposure of reinforcement wires from the hose material.
- Signs of kinked, cracked, crushed, flattened or twisted areas in the hose sections.
- Hose ends degraded, pitted or badly corroded at the fittings.
- Identify sufficient numbers of floatation collars are on the hose string.
- On completion of bunkering operations the hose should be re-examined for any damage that may have occurred during the transfer operation.

INSTALLATION PROCEDURES

- Documents should be in place clearly specifying how the site will control the maintenance and inspection of all bunkering hose strings and associated equipment i.e. lifting equipment and support mountings. This document should be approved by the relevant Technical Department and entered into their pertinent system for review as per the Company Procedures. The appointed system owner is responsible for ensuring all relevant persons know of and understand the procedure. It is recommended a competent/responsible person carries out frequent lifting equipment audits to confirm this.

RECOMMENDED CONTENT OF PROCEDURES

- The system owner should indicate who is responsible for ensuring the procedure is being adhered to and act as focal point on all matters relating to bulk hoses maintenance and inspection.

- On locations where bunkering of drilling products takes place, an interface should exist with the drilling department and operations departments where responsibilities are clearly defined, documented and agreed i.e. who is responsible for inspection and change-out of drilling product hose assemblies. The role of service team leaders, barge engineers or deck foreman should be considered for system owner positions.
- Guidance based on information gathered from the hose manufacturers on the life span of in-service hoses before mandatory change-out is required. Identify the time periods between physical and visual inspections including pre and post use checks of the equipment. This decision is addressed with relevant parties such as suppliers and company technical authorities. An Electronic Maintenance System (EMS) would be ideal to populate / generate change-out dates and inspection dates, and guidance on the required documents e.g. Permits to Work, COSHH and Method Statements to carry out the work scope safely. On completion of any parts being changed-out, documents and identification (ID) of equipment must be updated in the hose register. In the case of replacement hose assemblies already stored offshore, a guidance note on the correct procedure of storage and shelf life should be obtained from suppliers.
- To assist in the managing, ordering and replacement of bulk hose equipment, drawings which may be electronic or hard copy, consisting of the following, would ensure the correct parts are ordered and installed at all times:
- The correct hose lifters (hooky hooks) and their SWL.
- The type of delivery coupling, be it self-sealing or hammer lug unions.
- The correct type and quantity of floatation aids / collars and their positions in relation to the string and joining couplings.
- Describe the type and SWL of the hose lifting assembly used for transferring the hose string during operations.
- In the case of strings being made up from both hard and soft wall sections their chosen positions should be identified in the drawings.
- Identify all components by part numbers.
- Method Statements, Lifting Plans and Risk Assessments must be in place and available for the work party. The system owner and work party should review these documents before use, however if on completion of the task lessons were learned, the documents should be updated accordingly by identifying the changes in the procedures.

GOOD PRACTICES

There are alternative systems available such as portable saddles which support the arc of a hose when in storage. The structure from which the hose is to be suspended must be surveyed by a competent person to ensure the hang-off point is of sound structure. Examples of bad practices are included at the end of the annex.

HOSE COMPONENTS AND CONSTRUCTION

All new hose sections are hydro-tested to at least 1½ times their working pressure. A water hose is made from orange coloured, soft, reinforced rubber with the cover being made of ethylene propylene diene-terpolymer (EPDM) hose reinforcement being provided by multiple layers of rot-proof synthetic textile yarn. The central core/tube is made from non-toxic and non-tainting rubber. The cover is abrasion and weather resistant, and care should be taken when handling and stowing. It should be noted that new floating hoses are also coloured orange and these hoses can carry a range of products. A fuel hose is heavy and commonly soft wall type, but can be of hard wall construction. The outer wall is made of black oil resistant neoprene synthetic rubber and is reinforced with synthetic textile yarn with antistatic copper wire. It has a black nitrile tube. The outer cover on this hose is susceptible to mechanical damage. The hose carries a brown lazy spiral stripe for identification.

HOSE STRING

A hose string can be made up of 3 or 4 lengths of 15.2 metres (49.86 feet), 16.3 metres (53.4 feet) or 18.3 metres (60 feet) lengths of hose joined together by quick release self sealing couplings (hammer unions). The hose comes complete with a hose lifting assembly that consists of a hooky hook, lifting sling not less than two metres (6.5 feet) in length and a safety pin shackle. The pin used to secure the nut must be a "split pin" and not an "R" clip. "R" clips can spring off the pin affecting the security of the shackle. When ordering new hose sections stipulate the direction of the lifting eye, as the hooky hook can be installed on the hose with the lifting eye facing up or down on the hose. If the hose is stored in a support frame then the eye in the hooky hook should be facing upwards, if using any other type of hose support then the eye on the lifter can be either way on the hose.

Hose Lifter (commonly known as hooky hook). The hose should be fitted with the correct number of floatation collars to prevent the string sinking and being drawn into the supply vessel's thrusters. The floatation collars can also be used to help form a barrier between the hose and installation structure by simply adjusting the collar straps on the hose. Reflective floatation collars have an advantage when bulk is being transferred to an installation in the hours of darkness as the crew can see the hose is floating freely rather than being to close to the vessel side thrusters.

New Hose Storage. Hoses delivered to the installation are normally shrink wrapped and rolled up with one end of the connection in the middle of the roll. It is preferable to store these hoses flat, out of sunlight and free from water ingress. Ultraviolet radiation and kinking during storage may shorten the life span of the hose. The original equipment manufacturers' recommendations on hose storage should be available for crews to ensure optimum methods of use to prolong hose life.

REPLACING SECTIONS OF HOSES IN A STRING

Only competent personnel should carry out the installation of hoses and connections when joining hammer lug unions. When repairing a hose string the hose should be landed rather than left hanging from a crane hoist line. When replacing a section of hose it should be inserted in the coupling and secured whilst free from tension. Once the necessary controls such as permit, method statement and risk assessments are in place then remove and replace the worn parts of the string. When hammer lug unions are disturbed the unions should be tightened up and marked across the body with paint or a similar permanent marking. This is a simple way to indicate if the coupling has slackened off due to movement whilst in service. On completion of hose installation, the hammer lug union should be checked for the marks across the coupled joint to confirm security. If possible pressure test the hose string to 5 bar and check the assembly is free of leaks over a 5 minute period. Check the correct quantity of lace up or similar types of floatation collars are on the hoses in accordance with *Table P-1*.

Table N-1 Recommended Floatation Collars for Bunkering Hose Strings

Hose Application	Hose Diameter	Floats per Section	Colour Code	Connection
Potable Water	3"	4	Blue	3" and 4" Hammer Lug Union
	4"	4		
Oil Based Mud	3"	9	Red	4" Hammer Lug Union or self sealing hose connections
	4"	10		

Dry Cement	4" 5"	7 8	Yellow	5" Hammer Lug Union
Diesel Fuel	3" 4"	4 4	Brown	4" Self sealing hose connection
Dry Barite	4" 5"	10 13	Orange	5" Hammer Lug Union
Methanol	2" 3"	4 4	Purple	4" Self sealing hose connection
Drill Water	4"	4	Green	4" Hammer Lug or self sealing hose connection

The above information is a recommendation from North West European Area Guidelines. As a minimum requirement for best practice, a float either side of any coupling that is in the water during bunkering operations would be advised. Trials of "floating type" bulk hoses have recently been carried out on some installations and have proved to be very successful, with positive feedback from all concerned platform and supply vessel personnel. As a result of these trials a major operator decided that this type of hose shall be used in the future. Therefore whenever a bulk hose section has been deemed to be no longer fit for purpose, the replacement ordered shall be of the floating type if the section, when in use, will be floating in the water. It should be made clear at this point that we do not expect all installation bulk hoses to be changed out en masse but on an "as and when required" basis. Each installation shall consider that when a new drilling campaign is due to begin and the associated bulk hose sections are ordered, to only order the new type for sections that when in use will be floating in the water. The new type of hose shall be used for every type of bulk cargo transfer. The pre and post use inspection of the new type hoses carried out by the deck crew shall remain the same. An appointed contractor will receive instructions from the manufacturer on the inspection criteria to be carried out by them. The ideal make up of the bulk hose string shall be either 3 or 4 standard lengths (18.3 metres (60 feet) depending on the installation needs and the elevation of the manifold. There is no requirement to have the first section of bulk hose leading from the manifold and not coming into contact with water, to be of the floating type. A typical hose string of 3 lengths would be:

Length 1: Hard wall
Length 2: Floating hose
Length 3: Soft wall (outboard / vessel end)

WEEKLY INSPECTIONS

A regular inspection PMR / Work Order signifies a competent person has assessed the hose and lifting equipment and that it is in good working order. This person records the findings electronically in a controlled register. This system indicates to any 3rd party auditors that a sound maintenance strategy is in place to manage bulk hose assemblies. Check all lifting slings, shackles and hooky hooks are in good condition and display current lifting colour codes. Check the hose for any physical damage for chafing, cuts, blisters, splitting, perishing, lacerations or other forms of deterioration. Renew any damaged hoses in the string and where minor damage is evident record details on the check sheet. Check markings across the hammer lug union line up as this indicates the fitting is tight on the coupling. Check hoses are protected from platform structure and stowed properly in hang-off points. Check that hang-off point structures shows no sign of deflection or excessive corrosion. Consider inspecting hoses inboard once per trip as there are blind spots on the installation structure that restrict visual inspections. Check the under-deck lighting on the installation is operational at valve manifolds. Check gates on the bunkering station hang-off points (fingers) are lubricated and easy to open and close. A record of visual and routine inspections should be available for history and evidence of hose checks.

Table N-2 Record of Inspections on Bulk Hose Lifting Arrangements

RECORD OF INSPECTIONS ON BULK HOSE LIFTING ARRANGEMENTS					
LOCATION OF LIFTING EQUIPMENT	HOSE LIFTING SLINGS I/D NOS	HOSE LIFTING SHACKLES I/D NOS	HOSE LIFTING HOOKY HOOKS I/D NOS	CHECKED SIGNATURE DATE	COMMENTS

AFTER STORMS

A visual inspection should be carried out to confirm hoses show no signs of physical damage. Examples would be chafing, splitting, perishing or any other form of deterioration. It is not uncommon for hoses to become twisted around each other if they were not far enough out of the water when exposed to severe weather, making it an operationally difficult when realigning the bulk hoses.

VISUAL INSPECTION PRE AND POST USE

A visual inspection must be carried out prior to and after vessel operations. The following checks should be carried out as a minimum. Correct colour coded hooky hooks, slings and shackles with proper split pins are attached to the hose. Hoses must show no signs of physical damage to fabric by chafing, splitting, perishing, blistering, deep lacerations or any other forms of deterioration. Check installation manifold couplings are tight and ready for operation. When using certain types of hose fittings remove the end screw dust cap before lowering the hose to the vessel, and on return to the installation replace the dust cap and check it is secured to an anchor point. Check gates on the bunkering station hang-off points (fingers) are maintained, lubricated and easy to open and close. The preferred way to visually check a hose is to place oneself in a safe position at the hose station and direct the crane operator to slowly raise the hose, allowing you to visually inspect the hose for wear. Never allow the hose to be lifted close to the crane hoist rope safety cut-out. A similar method can be used to check the hose for damage when returning the hose to its hang-off point. *Note: take care to avoid the hose being lifted immediately over the head of the person doing the inspection.*

GENERAL PLATFORM ALARM (GPA)

If the installation goes to a GPA status then bunkering operations must cease. The supply vessel's master and crane operator must be notified immediately of the GPA and company specific procedures are then followed before reporting to their muster station.

VESSEL APPROACHING LOCATION

Before any operation involving bulk hoses is undertaken at installations, always refer to the following. The services supervisor or equivalent will review the weather, wind speed and sea state. The master of the vessel will review the operational limitations of his vessel and only when satisfied will enter the installation's 500 metre (1,640 feet) zone. Before the vessel arrives at the installation the master will discuss the relevant shipment with the services supervisor, services team leader (STL) or equivalent. The deck foreman will liaise with the crane operator and the vessel master to confirm if conditions are favourable for bunkering and transferring cargo. The installation and vessel must complete their respective pre 500 metre (1,640 feet) entry checks prior to the vessel entering the 500 metre (1,640 feet) zone. Once the vessel master is satisfied that he is on station the crane operator can then lower the hose to the vessel at a height that allows the crew to secure the hose to the vessel's side rail. The vessel crew need to be aware of the operation and maintain a line of sight with the crane driver on the installation to ensure that they do not stand under any suspended load. Once secure the hose end is lowered inboard of the rail and the crew disconnect the crane hook. Once the hook is clear the ship's crew will connect the hose to the appropriate manifold. The ship's crew should be reminded that the hose coupling should, wherever possible, avoid contact with the ship's structure and should monitor the integrity of the hose coupling during transfer. *Note:* In marginal weather great care is required by the vessel master to avoid over-running the hose especially if the cargo is also being transferred. Consideration should be given to the connecting of bulk hoses only at this time. During hose work deck foremen must listen to all communications on selected radio channels, which can be transmitted to the control room and platform crane operator should a hose assembly leak or significant changes in weather conditions occur.

POLLUTION SAFETY

During fuelling operations there is always the risk of pollution, this may be due to hose and/or instrument leaks, hose wear, mechanical breakdown or if a hose becomes fouled in the vessel's propeller. It is important that an individual is appointed to visually check and operationally check the hose remains functional during bunkering operations. If an oil sheen is detected on the surface of the water then bunkering operations must cease immediately. The incident must be reported to the control room and the cause investigated.

GUIDANCE ON BULK HOSE OPERATIONS

During bulk hose operations the following guidance should be observed. The vessel master, crane operator and deck crews to confirm radio communication prior to operations. The person appointed to supervise the bunkering process must ensure he can see the bulk hose(s) at all times, and that he is familiar with the alignment of valves and tank levels. He should not allow other distractions during the operation. The installation dry bulk vent line positions are identified. The vessel bridge or equivalent and OIM / barge master or equivalent should confirm quantities discharged and received at regular intervals, to ensure that there are no leaks within the respective systems. The vessel deck crew and installation crane operator must be readily available and close at hand throughout any transfer operation. Sufficient warning / instructions shall be given by each party prior to changing over the tanks. If at any stage in the operation the vessel master or provider is in any doubt as to the integrity of the operation, then that operation shall be suspended until integrity can be reinstated. When pumping is finished, both the receiver and provider shall set their line to allow the hose to be drained back to the vessel's tank. In suitable conditions the crane may also be used to lift the hose to aid draining. In the case of dry bulk, purge air should be used to empty the hose and clear the line. Hoses used for potable water must not be used for transferring other bulk liquids. Potable water lines should be flushed through prior to transferring water to avoid any residues within the lines contaminating the installation's supplies. During periods of darkness adequate lighting must be available over the hose and support vessel throughout the operation. To identify hoses they may be fitted with high visibility bands, tape or alternative means. Hoses are normally colour coded for manufacturers' identification and approval, frequently by way of spiral coloured bands within the hose structure. Ensure the management system is aware of the markings on the hoses. The manufacturers' colour coding of the hose should not be confused; any markings on receivers or structure should adopt the universal colour coding as described in *Annex K, Section 4.18* of the *North West European Area Guidelines* (NWEA) to identify bulk hose products. All bulk hoses used offshore are to be of sufficient length and good condition; unapproved repairs shall not be carried out, and in the interests of safety the hose should be disposed of immediately. In the event that the crane operator has to leave his cab, he should first inform the master of the vessel but must remain in radio contact so that he is on immediate call. Any bulk hose should be disconnected from the vessel as soon as possible after the bunkering has been completed and retrieved to the platform, unless other wise agreed by the master of the vessel. The vessel should ensure that all pollution prevention equipment is in place as per vessel's SOPEP. If a connection other than a self sealing quick release coupling is used, particular care must be exercised when disconnecting the load hose and a drip tray must be in place. All manifold valves have been checked and confirmed to be in good condition. Correct couplings have been identified for the product(s) to be transferred. The PIC of the operation performs no other duties during the transfer(s).

RECENT DEVELOPMENT IN HANDLING BULK HOSES

Introduction. The offshore oil and gas industry has identified a considerable rise in hose snagging incidents involving supply vessels, and enlisted the vessel masters to assist in the development of a safety system to minimise the risks when receiving and removing bulk hoses from and to supply vessels. As a result of this, a method has been developed which resulted in minimal modification to the ships and minimal physical handling of the hose.

Bulk hose hang-off sling attaching and inspection procedure. It has been identified that the use of an endless round web sling attached to the bulk hose and hung over a dedicated point on the supply vessel enhances the operation of passing the hose to the vessel from the installation. This process only relates to the attaching of the web sling to the bulk hose and the frequency of inspection. The requirement for this endless round sling will depend entirely on the facility of a suitable attachment point on the supply vessel being available. A dialogue between the Services Supervisor and each vessel shall take place to establish the requirement for this sling and the distance from the end of the hose to attach it.

Attaching the endless round sling to the bulk hose. An endless round sling of three tonnes SWL must be used. The sling shall be of three metres (9.84 feet) effective working length. The sling shall be signed out of the rigging loft and attached before each use, and detached and returned to the rigging loft for correct storage after every use. The sling will be attached to the bulk hose using the "double wrap and choke" method. The attachment point for the endless sling will be approximately seven metres (23 feet) from the end of the bulk hose offered to the supply vessel; this distance will be confirmed by the vessel master. The endless round sling must only be attached to the bulk hose by a competent Rigger or a competent slinger / load handler. Once in position it shall be secured by tie-wraps or light cord to prevent slippage/loosening of the sling.

Inspection and storage of the endless round sling. The LOLER Focal Point or a competent person must inspect the sling before and after each use to ensure it is still "fit for purpose". The sling is to be inspected to cover the following points as a minimum:

1. Check SWL.
2. Check colour code is current and ID Number is legible.
3. Check entire length for cuts, tears or chafing.
4. Check joint for burst stitching.
5. Check for chemical damage and heat damage.
6. Check there has been no ingress of foreign bodies into the fibres.

When checking the round sling, should any cuts be found in the outer protective cover then the sling should be condemned, i.e., DO NOT USE as the inner strength core may be damaged. When the bunkering operation is complete the round sling shall be removed and returned to the rigging loft for storage.

Requirement. This requires three pins reasonably spaced out on the upper rail or taff rail on each side of the vessel to be welded in place, adjacent to the bulk hose manifolds. These pins are used to hook the eye of an endless webbing strop on to a three tonne SWL and ca. 2-3 metres (6.5-10 feet) long webbing sling when attaching the hose to the vessel.

Method. The vessel master may ask for the sling on each hose to be adjusted for his manifold and hang off points prior to coming alongside. This may vary according to the distance from the hang-off position of the required product manifold on the vessel. Under instruction the crane operator will transfer the hose from the installation to the vessel in the normal fashion. During the lowering of the hose as the hang-off strop nears the vessel's side rail, the crew will retrieve the eye of the strop by hand, or if necessary by boat hook, and fit the eye of the sling over one of the pins. Care must be taken by the vessel crew to avoid positioning themselves under the suspended hose during this operation. The crane operator, upon instruction, continues to lower the hoist rope until the sling takes the weight of the hose, the vessel deck crew then signal him to lower the hose end into the safe haven where they unhook the hose end, allowing the crew the freedom to manoeuvre the hose end onto the manifold. On completion of transferring bulk the vessel deck crew drain the line and remove the manifold connection. The connection is moved away from the manifold by the crew prior to signalling the crane operator to lower his pennant to the deck crew. The hose end is attached to the crane hook via the lifting sling, and once everyone is in a safe position the crane operator is given the signal to raise the sling until the hose and hang-off strop are clear of the vessel. This modification eliminates unnecessary risk to crews when transferring the hose back to the installation. A final inspection should be carried out on the hose and lifting assemblies prior to and after use, recording findings on a check list.

BENEFITS

Securing the hose is simple and very effective in comparison to making the hose fast by lashing it to the ship's side rail. Crew exposure to a suspended load is vastly reduced and minimal. Fingers are not exposed to the same risk when lashing the hose. Passing the hose back is much safer, as personnel involvement after hooking the hose end on is virtually eliminated. Minimum alterations are required to operate this system. Floatation Hose Strings are available when bunkering to and from installations.

TRANSFER HOSES

Historically many installations preferred to work supply vessels with three hard wall sections of hose. This was until evidence indicated that when using hard wall hose strings considerable problems were caused for the supply vessel deck crews. The supply vessel crews found it was difficult to manipulate the hose into position when connecting hard wall hoses to vessels' manifolds. Other examples were found when the installation deck crew were repairing or making up these hose strings. Installation and vessel crews reviewed the hazards caused when using hard wall hoses and agreed the first and second section of the hose suspended from the installation manifold could be hard walled, and the last section which is offered to the supply vessel to be of soft wall material, which would help reduce the incidents that the deck crew were experiencing when making up hose strings, specifically when connecting the installation hose to the vessel manifold.

Fig.N-1 Permanent structure used to support bulk hoses forming a perfect arc which enhances the lifespan of the hose

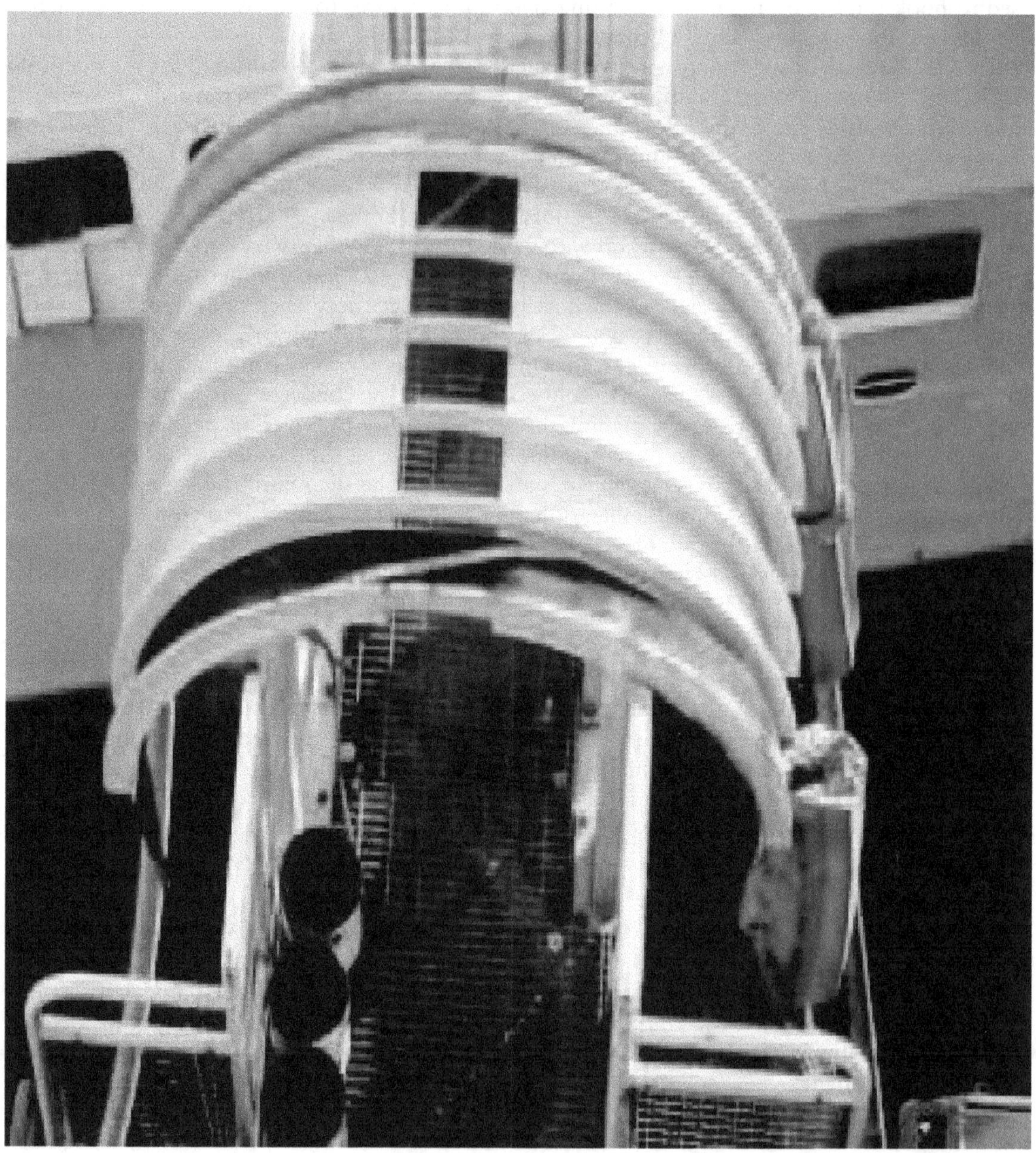

Fig.N-2 Hose lifter (commonly referred to as a hooky hook)

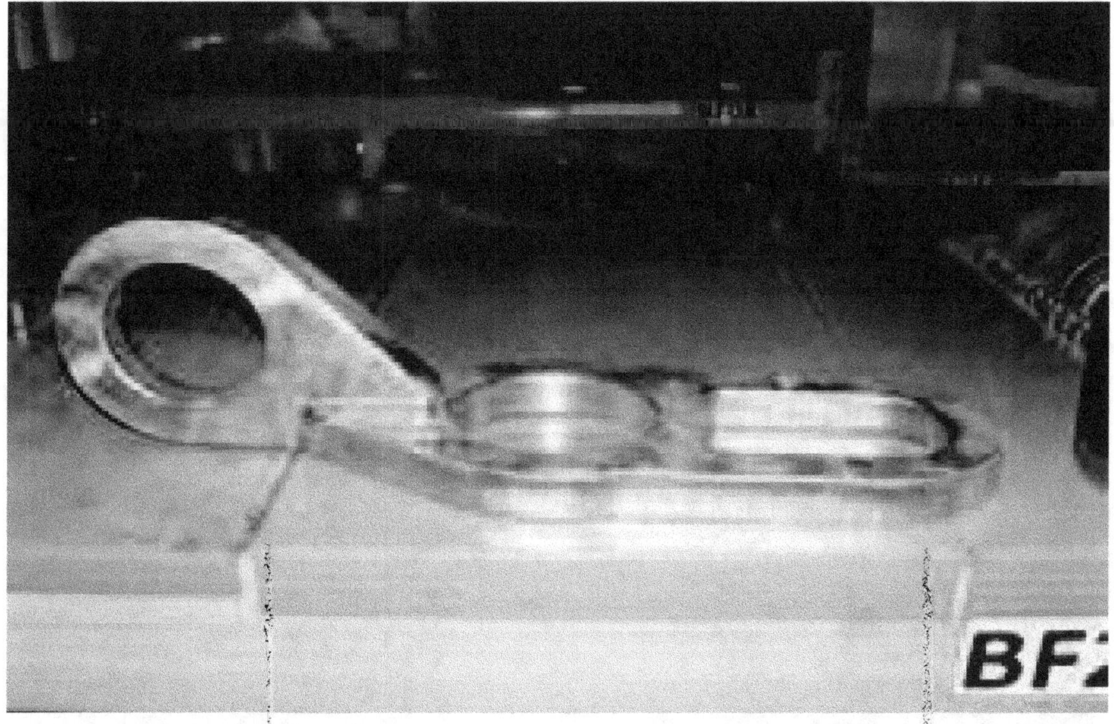

Fig.N-3 New Hose Storage

Fig.N-4 Attaching the Endless Round Sling to the Bulk Hose

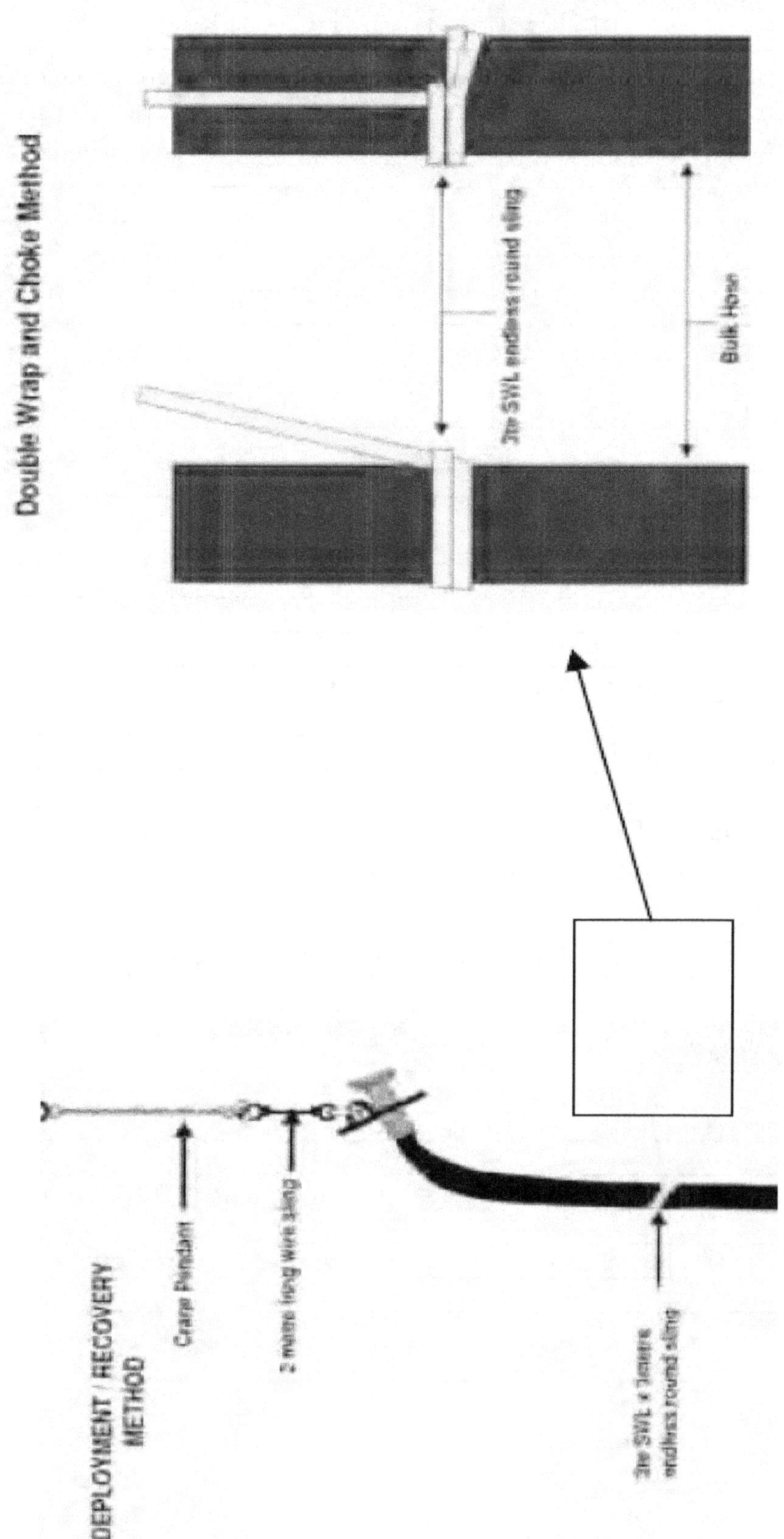

Fig.N-5 Pre Bunkering Check List

Pre-bunkering Checklist							Tick
Hose Inspected							
Diesel							
Pot Water							
Methanol							
Oil Based Mud							
Barite							
Dry Cement							
Drill Water							
Hose fabric is in good condition and shows no sign of perishing							
Hose end caps are fitted after bulk transfers							
Adequate floatation collars are fitted to hose string							
Hoses are suspended correctly and nor tangled							
Hose lifters (hooky hooks) are present, colour coded and fit for purpose							
Inspect the quick release coupling prior to and after use. Where available test against a blank coupling							
Hose lifting assembly is present, colour coded and SWL							
Hoses' protective covers preventing abrasion on structure are still serviceable							
This list is not exhaustive and can be developed for specific operations.							
Comments							
Checked/Signed							
Date							
Vessel Name							

Fig.N-6 Floatation Hose Strings are available when bunkering to/from installations

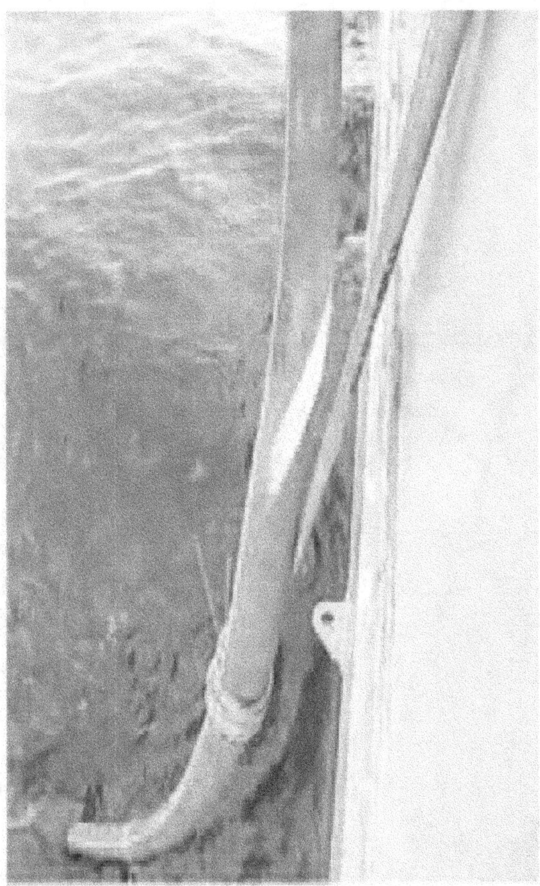

Fig.N-7 Floatation Hose Strings are available when bunkering to/from installations

Fig.N-8 *This eliminates the need to use floatation aids on the hose string.*

Fig.N-8 This eliminates the need to use floatation aids on the hose string.

Annex O

Bulk Hose Handling and Securing

An alternative method of handling bulk hoses to that described in *Annex P* is summarised below. This method also requires minimal modifications to the vessel and has been used satisfactorily in various areas. Modifications and preparations on the vessel include the following.

1. A rubber coating or similar arrangements should be installed on the cargo rails to provide friction so that movement in the hose(s) is prevented until secured to the manifold.
2. A sufficiently large area must be allocated and marked on the deck of the vessel so that the hose can be positioned by the crane without assistance from the vessel's deck crew.
3. Similar arrangements are required at all bulk handling stations where this method will be used.
4. The hose must have sufficient buoyancy elements, which must be clearly visible to vessel personnel.

In order to reduce the risks associated with bulk hose handling when using this method the following precautions should be observed:

1. A pre-job talk should be held between crane driver and vessel personnel.
2. The hose should be delivered with the crane hook connected to the end of the hose. Where this is not possible, i.e. where the hook is connected to the hose at some distance from the end, the free end must be secured to prevent uncontrolled movement.
3. Personnel on the deck of the vessel *must not* be in the allocated landing zone whilst the crane is handling the hose. After the hose is landed within the zone the crane hook is disconnected.
4. After the hook has been disconnected the hose is connected to the appropriate manifold prior to the commencement of the bulk transfer operation.

Hose Markings and Connections

HOSE MARKINGS

Hoses and hose terminations should be product-identified via high visibility bands, tape or other means. Below is the colour coding to be used for the Hose End Coupling (colour refers to coupling and not hose) which is passed to the supply vessel.

Table P-1 Hose Markings and Connections

Hose Application	Coupling Colour	Standard Connection	Vessel Coupling	Pressure Rating	Outline
Dry Cement	Yellow	5" hammer lug union	Male	Min 12 bar	
Dry Barytes & Bentonite	Orange	5" hammer lug union	Female	Min 12 bar	
Potable Water	Blue (Orange Hose)	4" hammer lug	Female	Min 12 bar	

Diesel / Marine Gas Oil	Brown	4" quick release self sealing coupling	Female	Min 12 bar	
Base Oil	White	4" quick release self sealing coupling	Female	Min 12 bar	
Drill Water	Green	4" hammer lug	Female	Min 12 bar	
Oil Based Mud	Black	4" quick release self sealing coupling	Male	Min 24 bar	
Brine	Red	4" quick release self sealing coupling	Male	Min 24 bar	
Glycol (if separated)	Purple	4" quick release self sealing coupling	Male	Min 12 bar	
Scale Inhibitor (if separated)	No colour	4" quick release self sealing coupling	Male	Min 12 bar	
Drill Cuttings (if separated)	No colour	5" hammer lug union	Male	Min 24 bar	
Methanol	Black and Yellow (tiger stripes)	4" quick release self sealing coupling	Male	Min 12 bar	
Water Based Mud (if separated)	Cyan	4" quick release self sealing coupling	Male	Min 24 bar	
Rig Slop (if separated)	Dark Gray	4" quick release self sealing coupling	Male	Min 24 bar	

NOTES

1. Manufacturers' identification or approval of hoses is often by spiral coloured bands. *Do not confuse* this with product colour markings.
2. References to figure values, in brackets, in the above table relate to particulars of the screw connection.
3. In some areas the same hose may be used for all classes of mud, being flushed between product changes.
4. Further information regarding hose types, testing, flotation, etc. can be found in Annex P.

HOSE CONNECTIONS

Some examples of hose connections are shown below.

Table P-2 Hose Markings and Connections

Description	Reference
4" quick release self-sealing coupling (Female)	
4" quick release self-sealing coupling (Male)	
Hammer lug (Weco) (Female)	
Hammer lug (Weco) (Male)	

NOTES

1. Self-sealing connections should be inspected when released to ensure full closure and that no liquid is being passed.

To accommodate possible mismatches, vessels should carry sufficient crossovers on board, typically including the following:

1. From 4" female hammer lug to 4" male quick release self-sealing coupling.
2. From 5" male hammer lug to 5" female hammer lug.
3. From 5" female hammer lug figure 50 to 4" female hammer lug.
4. From 4" male quick release self-sealing coupling to 4" female hammer lug.

Annex Q

Carriage of Oil Contaminated Cargo

OBJECTIVE

To provide specific advice for the safe transportation, offshore handling, tank cleaning, onshore handling and onshore disposal or treatment of wet bulk back-loads contaminated during drilling and other operations. This guidance is aimed at offshore installations, OSVs and appropriate onshore staff (for example, surveyors, tank cleaners, base operators, and waste processors). In particular, analytical tests must be carried out and made available to the ship's master prior to back-loading, confirming that flash point exceeds 60°C (140°F) and that the appropriate steps to avoid H_2S generation have been carried out.

BACKGROUND

Industry, in conjunction with the Chamber of Shipping and the Marine Safety Forum has produced this Good Practice document to assist operators in better describing the wet bulk back-load cargoes they wish to transfer to shore for processing, using the bulk mud tanks on OSVs. In the course of well operations, water based fluids such as seawater, brine or water based mud may become contaminated, commonly with oil based mud or base oil from oil based mud, (herein after called wet bulk waste) which cannot be legally discharged to the marine environment. These contaminated fluids are returned to shore for treatment or disposal. Operations giving rise to such fluids include:

- Well bore clean-up operations where oil base mud is displaced from the well bore to seawater or completion brine.
- Operations where water base mud becomes contaminated with oil base mud during displacements.
- Cementing operations with associated spacers.

- Pit cleaning operations.
- Drilling operations where well bore fluids are contaminated with oil based mud, crude oil, or condensate.
- Other tank cleaning operations where fluid chemical components cannot be discharged because of the Offshore Chemical Regulations.
- Rig floor drains where the fluid is oil contaminated.
- Any of the above fluids may also be contaminated with hydrogen sulphide (H_2S), typically from sulphate reducing bacteria (SRB) activity.

When fluids are severely contaminated and of small volume, then general industry practice is to transport to shore in Tote tanks or similar type carrying units. For fluids that are "lightly" contaminated, general industry practice has been to back-load to the mud tanks on the OSVs. It is this latter practice in particular that has raised grave concerns for the following reasons: (a) it is difficult to accurately describe the chemical make-up of the fluid and hence provide a MSDS sheet that adequately describes the material; (b) Gas testing on OSVs returning to shore with this cargo has found on a significant number of occasions, high levels of H_2S in the atmosphere above the cargo. Lower Explosive Limit (LEL) tests also revealed an explosive atmosphere in excess of that which the OSV has the capability to safely transport; and (c) the mud tanks on the OSVs are not designed or classified to contain and transport wet bulk cargo with a flash point of less than 60°C. The pump rooms and pumping systems for the discharge of the product tanks are not intrinsically safe. This classification is only found on board specialist type OSVs. The reason for the very high LEL % values that have been recorded is contamination with crude oil and condensate. The bulk mud tanks on standard OSVs are not designed for this purpose and under NO CIRCUMSTANCES should fluids contaminated with the mentioned products be back-loaded to an OSV's mud tanks.

Recognising the relatively complex nature of the cargo, this Good Practice document has addressed the issue by recognising that a series of tests should be undertaken on the material intended for back-load to provide an indicative view of the constituent make up and reactive qualities of the material. It must be recognised that because of the segregation issues described in section 3 of the Guidelines, these tests can only be indicative. The tests can be performed either on the rig or onshore but must be performed by a competent person as determined by the operator. The rate at which these fluids are generated during certain operations on the rig may preclude sending samples to shore for testing necessitating rig based testing. In either case, the results of the tests must be made available to the master of the OSV prior to the back-loading hose connection taking place. Once tests have been carried out no more fluid should be added to the intended cargo on the offshore installation. If any further additions are made a further test will be required. The results of these tests will allow the master to establish if the back-load is acceptable for carriage on board the OSV. Acceptance is based on the reported analytical information and the measured physical properties, the known nature of the chemical make up and the previous cargo carried in the OSV's tanks. A generic risk assessment will be available on board the OSV and updated when new, improved or different information and circumstances become apparent. Offshore installation staff should be aware that in certain circumstances the master of the OSV may require advice from the OSV's onshore technical advisors and that a response from onshore may take time to progress. If there is any doubt regarding results repeat the tests and review. The back-load hose should not be sent to the OSV and connected up unless there is an agreement between the OSV master and the installation OIM / operating diving support vessel that the back-load is acceptable for transportation.

COMPOSITION OF THE WET BULK WASTE

The final wet bulk waste may contain components and formulated mixtures including:

- Water (both seawater and potable water).
- Oil base mud.

- Base oil.
- Water base mud.
- Well bore clean up detergents.
- Completion brine (including corrosion inhibitors, biocides, etc.).
- Cement spacers.
- Rig wash.
- Brines containing various salts.
- Other substances, e.g., glycol, pipe dope, etc.

The major component is normally seawater. The proportions of the other constituents are variable. The wet bulk waste is likely to be heterogeneous in that oil mud will separate to the bottom, base oil to the top, with seawater in between. OSV motion will not normally be sufficient to mix and stabilise the cargo to a homogeneous form. The components and formulated mixtures may arise from different well bore operations. The volumes of each component are normally known, although the degree of volumetric accuracy is variable depending on how and where this material is stored on the rig prior to back-loading to the OSV. During discharge to onshore storage tanks and road tankers the make up of the initial discharge may be different in composition to that discharged later due to separation of components during transportation. This may result in higher concentrations of an individual component being transported in road tankers.

Example

Oil based mud or contaminated wet bulk waste containing:

Seawater	75%	(volumes)
Mineral oil base mud	10%	
Cement spacer with surfactants	10%	
Base oil	5%	

The above mixture will separate, leaving the base oil on the surface, the seawater below this and the mineral oil mud on the bottom. The cement spacer will mix with the seawater although the surfactants will also mix with base oil and oil mud. During transfer operations from the OSV to road tankers the initial fluid comprises the heavy oil mud, followed by the lighter seawater and finally the base oil. In the event of a hose rupture or spillage, all component fluids should be treated as oil contaminated and should be contained, preventing discharge to the sea.

TESTING PRIOR TO BACK-LOAD

A wet bulk waste may contain a significant number of chemicals for which MSDS are available offshore. It is not practicable, however, to develop a description of the wet bulk waste from such an array of documents. Although MSDS will be available for formulated mixtures, there may still be uncertainty in describing the properties of the wet bulk waste. As a precaution the following tests should be carried out, prior to back-loading, in order to assist confirmation of the potential hazards:

• pH	Numerical range	0 - 14
• Salinity (Chlorides)	mg / l	
• Retort	Oil content	volume %
	Water content	volume %
	Solids content	volume %
• Flash point	(closed cup °C)	
• Noxious gases	LEL Explosive gases	
	H_2S	
	Oxygen	
• Bulk density specific gravity		

As described in section 2 of the Guidelines, tests may be carried out offshore on the installation by trained and competent personnel or samples sent onshore for analysis by the waste processor or other competent laboratory. The analysis and treatment should be carried out in a timely fashion on representative samples of each wet bulk waste intended for back-loading to an OSV. If back-loading is delayed for any reason, such as bad weather, it should be noted on the analysis form attached as Annexe 10 - F - 2 to Appendix 10 - F and the volume and the pH of the Wet Bulk Waste should be monitored daily. If there is any doubt regarding results repeat the tests and review. Results of the tests along with the analyst's signature and date completed should be entered on the Annex 10 - F - 2 analysis form and attached to the appropriate Waste Consignment Note, e.g., SEPA C note.

Table Q-1 Key Test Results (Ranked)

Test	Indicator	Range of Results	Interpretation
Flash point	Potential for explosion	> 60°C	Should be > 60°C to back-load. If the flash point is low (<70°C) then an explanation should be provided.
LEL	Potential for explosion	Ideally zero. Meter alarm typically set to 10-20% LEL	Consistent with flash point above – for transport only. If measurable LEL, repeat test and review explanation..
H_sS	Poisonous gas	Must be zero.	Indication of bacterial activity.
pH	Measure of acidity or alkalinity	9.5 – 10.5	To keep H_2S in solution COSHH Personnel Protection Equipment and personnel exposure. If pH greater than 11, discuss with OSV master.
Oil % Volume	The major component requiring back-load	Agrees with components in Annex 10-F-2	Confirm retort agrees with Annex 10-F-2 and waste consignment note.
Solids % Volume	Potential need for tank cleaning	Agrees with components in Annex 10-F-2	Confirm retort agrees with Annex 10-F-2 components and waste consignment note. Tank residue could form a source of SRB and H_2S over time..

More detailed Procedures are provided in *Annex Q*, which is attached to Appendix 10-F of the Guidelines. Test results should be consistent with the information on the Annex 10-F-2 analysis form.

FURTHER GUIDANCE FOR THE OSV

No Wet Bulk Waste should be back-loaded until an *Annex Q analysis form* has been received on board confirming that it is acceptable for transportation. There is no onus on the OSV to carry out further tests. Tank hatches should not be removed offshore because of associated risks to vessel and personnel Tests on board the OSV at the time of back-loading are only possible if sampling ports are available. Consideration should be given to installing suitable sampling ports on board OSV's to allow the use of the LEL / H_2S meter. (Usually this can be dropped from the vent system using the extended sniffer hose). Loading on top of bulk fluids already in ships tanks should be avoided. Wet Bulk Waste should where possible be back-loaded to a suitable clean tank. Where this is not possible further guidance should be sought from operator's competent person and with reference to operator's procedures. The potential for biological activity resulting in H_2S in the dead volume and sludge must be risk assessed. Should the overall pH be reduced through mixing of the fluids H_2S breakout can occur. Wet Bulk Waste should be discharged from the OSV as soon as possible. The need to clean the tanks should be reviewed on each trip to minimise the risk of biological activity and H_2S build up from any solid residue. Experience has shown that round tripping untreated Wet Bulk Waste increases the risk of H_2S breakout occurring due to the additional time Sulphate Reducing Bacteria (SRBs) have to be active. *Important:*

Where Wet Bulk Waste is to be round tripped a sample should be obtained from the tank and the pH checked to ensure no change has occurred since analysis. The volume of Wet Bulk Waste should also be checked to determine if any ingress has occurred (seawater ingress into the tank will reduce the pH and introduce a food source for bacteria) Where a change has occurred further guidance should be sought from operators competent person and with reference to operators procedures.

TESTING IN THE HARBOUR PRIOR TO OFFLOADING

A gas test for LEL and H_2S must always be performed on the OSV tanks containing the back-loaded material prior to offloading in port as a matter of standard procedure. Waste processors should also check the Annex 10-F-2 analysis form parameters onshore. Prior to discharge, the ullage air space in the tank will be sampled by the waste processor, preferably in conjunction with the Surveyor, for LEL and H_2S, to confirm that no change of condition has occurred. Undertaking these tests will confirm that the Wet Bulk Waste is safe to offload. A sample from the offloaded material should be taken and compared to the original analysis. In the event that there is a significant divergence between offshore analysis and onshore analysis, the waste processor should raise a non-conformance. If there is any doubt regarding results repeat the tests and review. The offshore operator, the offshore location, the OSV master, base operator, surveyor, and tank cleaners should be advised accordingly.

Note: If the wet bulk waste is back-loaded into tanks already containing oil based mud residues as can be the case, then the onshore test results will be different to those measured on the rig.

DOCUMENTATION AND REPORTING REQUIREMENTS

MSDS documentation of the components and mixtures must be made available to the OSV master. IMDG manuals are carried on the OSV for all types of chemical materials shipped. A Waste consignment note appropriate to the area of operations, e.g. EA or SEPA C is generated to accompany the wet bulk waste being back-loaded. This should reference the attached Annex 10-F-2 analysis form. The completed Annex 10-F-2 analysis form is reviewed signed and dated by the operators representative to confirm the back-load is safe to transfer. The Waste Consignment note along with duly completed, signed and dated Annex 10-F-2 analysis form is to be made available to the ship's master prior to back-load operations for review and comment. Once it is agreed to back-load, a copy is forwarded to the waste processor onshore by the offshore Installation which will include volume of Wet Bulk Waste and estimated time of arrival in port. This will allow planning to ensure in most cases the Wet Bulk Waste is discharged in a timely and efficient manner reducing delays in port and likelihood of round tripping. A dangerous goods certificate must be provided by the offshore installation based on the requirements of the individual component MSDS. The Waste Processor checks the samples drawn onshore, comparing the analytical results to those obtained from the offshore analysis. In the event of a discrepancy the offshore operator, the offshore location, the OSV master, base operator, surveyor, and tank cleaners should be advised accordingly. Test results should be also be provided to tank cleaning companies in the event tank cleaning is required Whilst every effort has been made to ensure the accuracy of the information contained in this Appendix and its Annexes, neither, the Chamber of Shipping nor the Marine Safety Forum nor any of their member companies will assume liability for any use made thereof.

FLASH POINT

The minimum acceptable flash point (Pensky Martin Closed Cup or equivalent) of 60°C (140°F) is applicable to wet bulk wastes and will determine whether the material is safe for transportation via the OSV's tanks. SOLAS regulations determine that materials with a flash point below 60°C (140°F) cannot be back-loaded to a OSVs mud tanks unless the OSV is certified for carriage where additional systems of inerting the environment on board the OSV will be in place. Generally, OSVs do not have the intrinsically safe systems required for the carriage of produced or unrefined hydrocarbons. Sampling should be set up to detect the worst case situation, particularly where there is potential for crude oil or condensate contamination where the oil will rise to the surface of the tank. Drilling rigs will normally have robust ventilation in the area used to store oil contaminated fluids and this may mask the condition experienced on board an OSV when carrying hydrocarbon contaminated product. OSV storage tanks are not normally vented. Air sampling from above the drilling rig mud pits may understate explosive gases. Sampling should reflect the conditions in the OSV tanks i.e. no agitation. Base oils typically have flash points in the range 70 - 100°C (158°F - 212°F). If the only oil component in a bulk waste is base oil then the flash point cannot be lower than that of the base oil itself. If the flash point is relatively low (60 -70°C (140°F - 158°F) an explanation must be provided on the Annex 10-F-2 before the form is presented to the OSV master. Prior to sampling, the installation pit should be left without agitation for at least 30 minutes and then surface sampled. If there is any doubt regarding results repeat the tests and review. This sample can then be split and one part used for flash point testing and the other for noxious gases. Flash point is tested as per closed cup flash point equipment manufacturers instructions.

LOWER EXPLOSIVE LIMIT (LEL)

The LEL gas detector will confirm potential flash point problems. Note that the LEL meter is used in harbour to check vapour condition in the ullage air space above the tank prior to discharge. The test carried out prior to back-loading should reflect the conditions in the ships tanks i.e. there will be no agitation and no forced ventilation unless it is specifically required or requested (unlike rig mud pits). The Noxious gas test is modified to simulate the unvented ships tanks. The sample is placed in a closed container with a sampling port on top and left to equilibrate for 30 minutes. A tube is then connected from the port to the gas analyser and the sample analysed. This method simulates the unvented ships tank. The above Procedure has been agreed with gas analyser manufacturers and Service companies carrying out the test offshore. The flash point and LEL results should be consistent with each other. LEL gas meters are normally set so that the alarm goes off in the range 10 - 20% LEL methane equivalent. Any number above 25% would be considered high. Other gases potentially present can have a different LEL range than methane. If there is any doubt regarding results repeat the tests and review.

HYDROGEN SULPHIDE (H_2S)

H_2S can occur in well bore fluids but this source would normally be identified by rig equipment and appropriate measures taken to neutralise and remove the H_2S. In surface tanks and facilities H_2S most commonly arises from the activity of sulphate reducing bacteria (SRB). SRB will become active provided there is a "food" source and low oxygen conditions. This would be typical of stagnant oil contaminated fluid stored for a long time. This environment can arise on both installations and OSVs in tanks and manifolds. Disturbing stagnant fluids or mixing low pH fluid into a high pH fluid containing H_2S could cause the release of H_2S into the void space above the tank. Hydrogen Sulphide is a heavier than air and an extremely poisonous gas. Maximum exposure limit is 10 ppm over an eight hour period. The LEL gas meters currently being used also tests for the presence of H_2S. H_2S is a known danger during drilling operations. Offshore sensors and routine offshore analysis methods will detect if H_2S is a potential problem in bulk waste back-loads. In the event of a positive test another sample should be collected to confirm the result. If this second result is positive further work may be required to determine the source of the H_2S. A test using a Garrett Gas train (if available) will determine the levels of H_2S dissolved in the liquid. The SRB organisms thrive in a pH range of 5.5 - 8.0. The lower the pH the greater the breakout of H_2S. The back-load MUST be treated on the installation to prevent breakout of H_2S in the OSV tanks. Biocides kill the bacteria but do not remove dissolved H_2S.

H_2S scavengers will remove dissolved H_2S but do not stop biological activity. Caustic soda (or similar alkaline materials) will raise the pH and prevent H_2S gas breakout. In the event H_2S is detected, tests should be carried out offshore to determine the best treatment prior to back-loading. If H_2S is detected but no H_2S scavenger is added to remove the dissolved H_2S, this should be noted in the conclusions section of the Annex 10-F-2 analysis form. After treatment a final head space H_2S test should be carried out to confirm zero H_2S and noted on the Annex 10-F-2 analysis form before the hose is connected to the OSV for back-load.

Example Procedure for LEL% and H_2S meter only

Collection of Sample

The sample should be taken from below the surface of the unagitated tank to simulate the unagitated OSV tank. Most oil will be in the top layer and will give a worst case oil content.

1. Leave tank or pit unagitated for 30 minutes before taking a 2.5 litre sample.
2. Fill the sample into container provided, up to the marked line and replace screw cap lid.
3. If a magnetic stirred is available, mix for one hour before proceeding to gas detection. Two large magnetic fleas included in kit.

Gas Detection (% LEL value, combustible gases)

1. Ensure batteries have been fully charged. If not, place in charger and allow charging for 12 hours.
2. Switch instrument on in a clean air environment.
3. The detector will beep and run a set of self-checks once these are complete the screen will display three levels on the screen:

   H_2S: 000 ppm
   O_2: 20.9 %
   LEL: 000 %

4. The pump automatically starts and continues to run until the unit is switched off.
5. Remove the plugs in the sample container lid and place the sampling hose into the head space.
6. Any combustible gas will be registered on the LEL monitor.
7. After five minutes remove the hose and switch detector off by holding down the on / off button for five seconds; (the unit will beep four times before switching off).
8. Any gases detected should be reported on the Annex 10-F-2.

Calibration

1. O_2 sensor is automatically calibrated each time the unit is switched on.
2. LEL sensor is factory calibrated to Methane and can be calibrated using a calibration gas supplied by BW Technologies.
3. H_2S sensor is factory calibrated but subsequent calibrations can be done using a calibration gas supplied by BW Technologies.
4. It is recommended that the LEL and H_2S sensors be calibrated every three months or when the unit is on shore using the appropriate mixed calibration gas from BW Technologies.

Sampling of liquid and solid component properties

The sample for the following analyses should be taken from the middle of the pit immediately after adequate agitation.

pH

Seawater pH is typically 8.3. Oil mud is alkaline and could raise the pH slightly. Cement contaminant is highly alkaline. In general alkaline pH (above 7) protects from corrosion. Highly alkaline materials can be caustic and require care in handling. Cement and sodium silicate can lead to high pH. Low pH (less than 4) is highly acidic and an explanation should be provided on the Annex 10 - F – 2 analysis form. Acids such as citric acid or acidising chemicals such as hydrochloric acid can lead to low pH. Low pH Wet Bulk Waste is very uncommon and would require large quantities of alkaline material to increase pH above 9.5. In this unlikely event further guidance should be sought from operator's competent person. Note that low pH (less than 9) means any H_2S present will already have broken out as a gas. The pH range of 4 – 11 is the acceptable range for transportation of any bulk fluids to avoid damage to OSV tank coatings and seals. Some OSV tanks may be capable of carrying fluids out with this pH range; this should be discussed with the OSV master prior to back-loading. Wet Bulk Waste will be treated to have a pH of 9.5 – 10.5 as this is the range that H_2S will remain in solution.

Salinity – Chlorides

Seawater is typically 20,500 mg / l chlorides. Oil mud contains some calcium chloride increasing this level slightly. Sodium chloride brine can contain up to 189000 mg / l. Results should agree with the composition.

Retort analysis (solids, water, oil volume %)

This should match the estimated composition (volume %) on the Annex 10 - F - 2 analysis form. Note: that it may be difficult to get representative samples if the liquid tends to separate. Some divergence is expected e g. if oil is noted as 5%, the range could be 3 - 10%. If separation is likely a range is preferred, e.g., 5 - 10%. The solids component can form a residue in the OSV tank and a potential location for SRB activity and H_2S.

SPECIFIC GRAVITY

Common water based fluids cover the range 1.03 (seawater), sodium chloride (1.2), and calcium chloride (1.33). Rarely used brines such as caesium formate can reach 2.2. Oil mud is typically 1.1 - 1.5 but can exceed 2.0. Mixtures will have intermediate values, most tending to 1.03 as seawater is the major component. Note that if mixtures separate the top half can be a different density than the bottom half.

APPEARANCE

General description confirming if cloudy, clear and colour. Should be consistent with Waste Consignment Note description.

ODOUR

Slight versus strong odour, consistent with description.

CONCLUSIONS

Should demonstrate the various parameters measured are in agreement with one another.

ANALYSIS FORM

TO BE COMPLETED AND PROVIDED TO OSV MASTER PRIOR TO BACK LOADING NO SECTIONS TO BE MARKED N/A			
Sample Description		Sample Reference	
Vessel		Date	
Offshore Asset		Producer	
Well Name and Nos		Waste Company	
Total Nos of Barrels		Waste Note Nos	

WASTE COMPONENTS			
Component Name	**Concentration**	**Units**	**MSDS Available**
		% Volume	
		% Volume	
		% Volume	
		% Volume	
		% Volume	
		% Volume	
		% Volume	
		% Volume	

LABORATORY ANALYSIS RESULTS				
Test	**Method**	**Units**	**Results**	**Range of Results**
Salinity (chloride)	Titration	mg/l		
Flashpoint (oil fraction)	Closed Cup Flashpoint	°C		Must be **>60°C** to back load. If the flash point is low (**<70°C**) then an explanation should be provided.
Gas Test (H₂S)		ppm		**Must be Zero**. Indication of bacterial activity.
Gas Test (LEL)	Gas Meter	%		**<25%**. Ideally zero. Meter alarm typically set to 10-20% LEL. Should be consistent with flashpoint.
Gas Test (Oxygen)		%		
pH	pH Meter			4 - 11 is the acceptable range for OSV tank coatings, **MUST** be 9.5 - 10.5 to keep any H₂S in solution.
Water	Retort	% Volume		

Fig.Q-1 Example of the Annex 10-F-2 Analysis Form

Oil Content	Retort	% Volume		Confirm retort agrees with Appendix 10-F, Section 4 components and waste consignment note.
Solids	Retort	% Volume		Confirm retort agrees with appendix 10-F, Section 4 components and waste consignment note.
Bulk Specific Gravity		SG		**<2.5**. If >2.5 seek further guidance on vessel capability.
Appearance				
Odour				
Date and Time of Analysis				

Conclusions:

Analysis to be carried out by a person competent to do so:	Comment (Yes / No / Details)
This liquid has been analysed as per GOMO Appendix 10 - F and it is my opinion that it is safe for carriage in a standard clean OSV bulk tank.	
This liquid has been analysed as per GOMO Appendix 10 - F and will be loaded into a tank with residues/existing cargo. Compatibility has been risk assessed and found to be safe for carriage.	
H₂S Avoidance	
Details of mandatory wet bulk waste treatment with biocide (chemical / qty)	
Details of wet bulk waste treatment in order to produce a pH of between 9.5 and 10.5 (chemical / qty)	
Has waste handling facility been informed of volume and ETA onshore? (yes / no)	
Does waste handling facility have the capability to take off waste at first port call? (yes / no)	

	Name	**Signature**	**Date**
Analyst			
Operator's Representative			

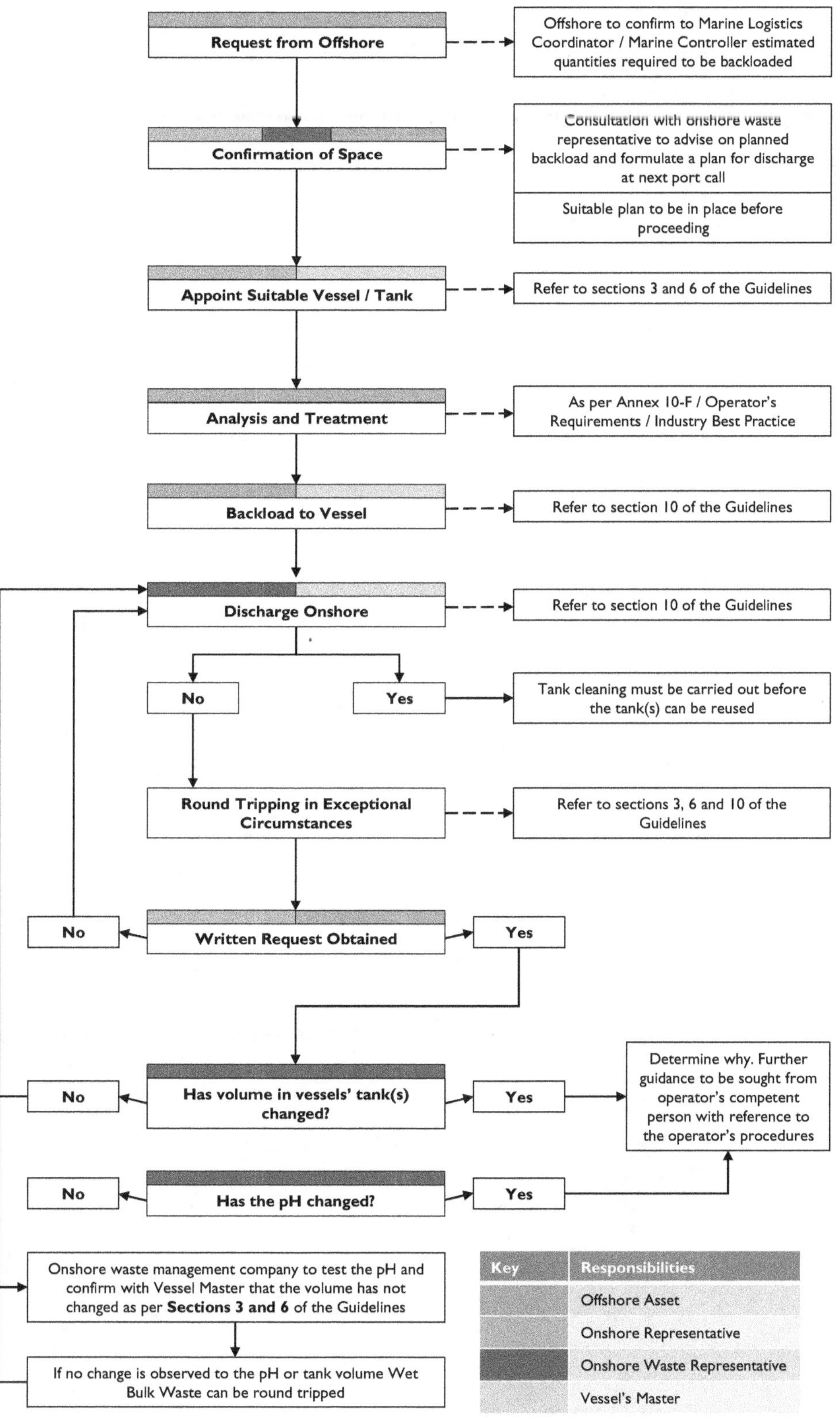

Fig.Q-2 Process Flow Chart

Guidelines for the MOU Move and Anchor Handling Work Specification

The following information is adapted from the Guidelines for the Content of MOU Move and Anchor Handling Work Specifications, issued by the Marine Safety Forum, dated April 2012.

GENERAL

Following onto the Bourbon Dolphin casualty the Marine Safety Forum formed several working groups to review the procedures and working practises that were currently in place for MOU move operations. These guidelines should be read in conjunction with the North West European Area Guidelines for the Safe Management of Offshore Supply and Anchor Handling Operations. This document is a Guideline to assist in producing an industry standard format for MOU Move and Anchor Handling Work Specifications.

DOCUMENT CONTENTS

Front Cover Sheet

Details of procedure author, checking process, revision history and dates; and approval names and signatures.

Table of Contents and Abbreviations Used

Introduction

To be specific to the operation and contain operational summary

- Detail requirements for Offshore pre-meeting with all vessel masters in attendance via con-

ference call to discuss Risk Assessments.
- Time Break Down Estimates.
- It is recommended that key personnel such as tow master, MOU PIC and operators representative sign to the effect that they agree to follow the MOU Move Plan.
- Also that at the "Hold Points" these same key personnel collectively agreed that conditions are suitable to commence the next phase of the operation.
- Limiting Environmental Criteria.
- Overview of Notification and advisory requirements.
- List of supporting documentation such as:

   1. MOU operations manual.
   2. MOU safety case.
   3. MOU owner and operator references.
   4. Local and regulatory Guidance.
   5. North West European Area Guidelines for the Safe Management of Offshore Supply and Anchor Handling Operations (NWEA) latest revision.
   6. HSE Operations Notice(s) 3, 6, and 65.
   7. HSE OSD 21 for Jack Ups.
   8. Warranty Certificate of Towage Approval.

Define **"Hold Points"** and **"Trigger Points"** to be used. Hold points are phases of the operation which may be completed in isolation or in a limited weather window. Hold points should be at suitable breaks which prompt a discussion to determine if it is appropriate to commence a phase of the operation. Examples include:

- Prior to recovering Secondary Moorings.
- Prior to recovering Primary Moorings and Going On Tow.
- Prior to Entering 500 metre (1,640 feet) zone at destination location.
- Prior to manoeuvring / mooring operations at destination location.
- Prior to running Secondary moorings.
- Prior to moving along side or over another structure.
- Prior to jacking operations.

Trigger points are defined as occurrences or events which would trigger a discussion as to whether it is safe to continue with the current operation. Examples include:

- Significant sea height reaches *xx* metres.
- Wind speed reaches *xx* knots.
- Current reaches *xx* knots or adversely impacts on operations.
- Visibility reduced to under one mile (1,760 yards).
- Unexpected loads experienced at either by any Anchor handling vessel or the MOU.
- Mooring equipment problems.
- Any mechanical problems aboard any Anchor handling vessel or MOU which may affect the operation.
- Any technical faults with the survey and navigation equipment.
- If at any stage there is any doubt about being able to maintain the clearance between the chains and wires and any sub sea asset.

Health Safety and Environmental

Should include a statement of Health, Safety and Environmental expectations for the MOU Move Operation and define which policies are to be used. Should include that all personnel are empowered to intervene; if for any reason the feel it is unsafe to continue, they do not understand or know what to do, or the agreed plan is not being followed:

- Outline of pre move meetings and briefings.
- Define Management of Change process.
- Define the Risk Assessment processes to be used offshore.
- Define vessel stability and loading requirement s for supporting vessels.
- Conduct a Post Move Review.
- It is recommended that on completion of a MOU move, the key players and vessel masters conduct an after action review to evaluate what went well and to identify possible areas for improvement.

Description of MOU

- Unit type, i.e. Semi Submersible, Jack Up, Barge, FPSO.
- Critical dimensions and key information taken from MOU Operations Manual.
- Mooring size, type and length, system of numbering of anchor lines and anchor patterns for both departure and arrival locations.
- Anchor types, weight and quantity.
- Towing gear arrangement Maximum Working Load, based on percentage of Maximum Break Load of weakest component.
- Propulsion systems (size and type of thrusters, DP on MOU / MODU).
- Draft and free board at both locations and during the tow. This should include such as jack up leg protrusion below the hull.
- Where appropriate the MOU data card to be made available to vessel masters.
- Detail any environmental limits for MOU operations.

Support

- Vessel requirements shall be based on defined expected worst case dynamic loadings.
- Mooring and rigging equipment.
- Towing Arrangements.
- Survey and Navigation package.
- Weather forecasting service.
- Tidal and current information.

Departure Location

- Positions / coordinates.
- To contain topographical diagrams of sea bed showing sub sea structures, pipelines and obstructions, slopes and such as sand hills and shallows.
- Drawings of current and proposed anchor patterns showing all mooring arrangements.
- Schedule for mooring recovery.
- Details of any skidding operations within mooring patterns.
- Water depths.
- Bottom type.
- Catenary curves.
- Soils data and penetration curves for Jack Ups.
- Leg extraction procedures for Jack Ups.
- Positioning tolerances and closest points of approach.
- Confirm minimum distances horizontal/vertical to installations and pipelines for anchors and mooring lines, including elevated catenary.

Towing

- Expected Duration of tow and Distances involved.
- Adverse Weather contingencies and safe havens.
- Emergency procedures.
- Responsibilities and command of tow and any transfer of that command.

- Passage Plan, Routing.
- Towing Catenary details or restrictions.

Arrival Location(s)

- Positions / coordinates.
- To contain topographical diagrams of sea bed showing sub sea structures, pipelines and obstructions, slopes and such as sand hills and shallows.
- Drawings of proposed anchor patterns showing all mooring arrangements.
- Schedule for mooring deployment.
- Details of any skidding operations within mooring patterns.
- Water depths.
- Bottom type.
- Catenary curves.
- Soils data and penetration curves for Jack Ups.
- Pre-load requirements for Jack Ups.
- Positioning tolerances and closest points of approach.
- Cross tensioning of moorings at arrival location.
- Contingencies for mooring slippage.
- Confirm minimum distances horizontal/vertical to installations and pipelines for anchors and anchor lines, including elevated catenary.
- On Shore Planning.
- On shore MOU Move Meeting to be held where operational procedures and responsibilities will be reviewed and agreed upon.
- Carry out an on shore Risk Assessment.
- Identify who hires and selects support vessels, mooring and towing equipment, navigational equipment, etc.
- Specify details of additional mooring, towing and navigational equipment required in support of the MOU move and Anchor Handling process.
- Identify who will brief vessels.

Refer to MSF 'Risk Assessment Rig Move Operations'. The MSF 'Vessel Health, Safety and Environmental Check List' to be completed at this time.

- Identify who will brief the tow masters.
- Refer to MSF 'Risk Assessment Rig Move Operations'.
- Identify who will carry out mooring analysis, to include anchor pattern, step / skid drawings, catenary curves. Details drawings to be included in procedure.
- Determine who provides Weather/wave data and tidal streams.
- Identify SIMOPS that may affect MOU Move operation and define who has primacy
- Define shut downs or bleed down operations required of adjacent installations and / or pipelines.
- Determine logistics requirements and responsibilities to ensure all equipment is sourced, certified and shipped.
- Any change of hire details during move are defined.
- Agree roles and responsibilities for OIM / PIC, barge master, tow masters, operator's marine representative, warranty representative, vessel masters, etc.
- Appoint sole point of contact through which all rig move notifications and exterior communications will pass.
- Agree 'Contingency Planning' requirements.

Appendices

Items for inclusion:

- Roles and Responsibilities for key roles such as OIM / PIC, tow masters, operator's marine representatives, positioning and navigation providers.
- Names and contact details of all parties, including emergency contacts and onshore support.
- Rig Move Risk Assessment, e.g., MSF Level 1 HIRA.
- Regulatory or local requirements.
- Tow route and passage plan.
- MOU data card.
- Drawings, charts and sketches.
- Description of anchor types and the use of mooring configuration drawings.
- Additional equipment.
- Sea bed bottom surveys and bathymetry.
- Tidal and current information.
- Jack Up Leg Penetration Charts.
- Location and warranty approvals.
- Cross track limits for deploying and recovery of moorings.

Cross tracking limits shall define the allowable vessel deviation off the intended mooring line, considering the distance from the MOU and amount of mooring deployed. A "Traffic Light" alert system is recommended.

Table R-1 Recommended 'traffic light system'

Zone	Limits	Action
Green	> 150 metres each side of intended track	No action required
Amber	Green zone + 150 metres on each side of the intended track.	Vessel instructed to regain line, assistance from rig provided if required. Review environmental forces being experienced
Red	Out with green or amber zones.	Mooring operation suspended until vessel regains amber zone and movement toward intended track confirmed. PIC notified.

Adverse Weather Criteria, Response and Rescue

ENVIRONMENTAL CONDITIONS					TYPICAL RESPONSE & RESCUE OPERATIONAL LIMITS			
TYPICAL WIND SPEED RANGES Knots (Metres per Second)			TYPICAL WAVE HEIGHTS		SIGNIFICANT WAVE HEIGHTS (H_s)	OPERATIONS INVOLVED		
ACTUAL SPEED at 10 METRES	EQUIVALENT AT ELEVATIONS		SIGNIFICANT (RANGE)	MAXIMUM (TYPICAL)		SBV SUPPORT	AVIATION SUPPORT	WORK OUTWITH PERIMETER
	50 metres	100 metres						
0 ~ 30 (0 ~ 15)	0 ~ 33 (0 ~ 16)	0 ~ 34 (0 ~ 17)	0 ~ 3.0	4.0	3.5	Limit for normal FRC deployment & recovery		Suspend work in orderly manner
30 ~ 40 (15 ~ 20)	33 ~ 44 (16 ~ 22)	34 ~ 45 (17 ~ 23)	3.0 ~ 5.5	7.5	5.5	Limit for emergency FRC deployment & recovery	No start-ups above 45 knots (22.5 m/s) at helideck	
40 ~ 50 (20 ~ 25)	44 ~ 55 (22 ~ 27)	45 ~ 57 (23 ~ 28)	5.5 ~ 7.0	10	7.0	Limit for use of mechanical means of recovery		
50 < (25<)	55 < (27<)	57< (28<)	7.0 <	10 <		No longer good prospect of recovery from sea. Vessel's own safety takes precedence.	60< knots (30< m/s) at helideck H_s 7< metres Routine helicopter operations suspended	

Wind Speed in Knots
Wind Speed in (Metres per Second)

NOTES

1. Wind speed is in knots. Wind speed in metres per second in brackets

2. Conversions between knots and metres per second are approximate only.

3. Elevations are above mean sea level.

4. Heights are in metres

5. Increase in wind speed with height calculated using ISO recommendations.

6. Assessment of conditions should include use of calibrated fixed or hand-held anemometers together with consideration of present and forecast weather.

7. Criteria relating to roll, pitch and heave of the helideck on floating facilities should be established by the aircraft operator.

8. The facility manager, in consultation with the ERRV Master, will decide when work outwith the perimeter should be suspended.

9. The facility manager, in consultation with the ERRV Master, HLO and aircraft Commander, will decide when flying operations should be suspended.

10. The facility manager, in consultation with the ERRV Master and / or aircraft Commander as relevant, will decide whether operations should be suspended in any other circumstances involving adverse weather, including reduced visibility, icing, etc.

Flowcharts illustrating the Handling of Bulk Cargoes in Port and at the Installation

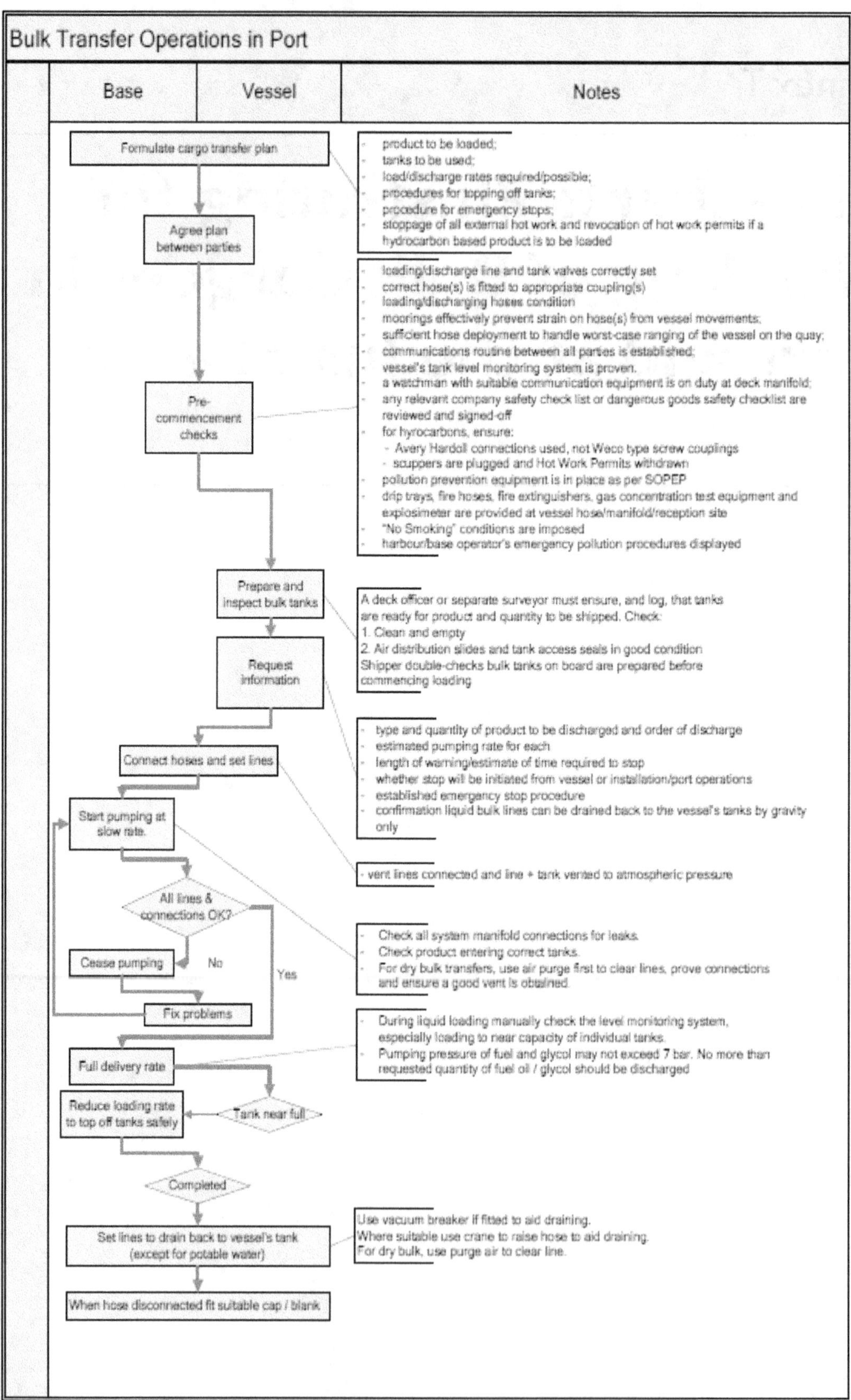
Bulk Transfer Operations in Port

Base
Vessel
Notes

Formulate cargo transfer plan
- product to be loaded;
- tanks to be used;
- load/discharge rates required/possible;
- procedures for topping off tanks;
- procedure for emergency stops;
- stoppage of all external hot work and revocation of hot work permits if a hydrocarbon based product is to be loaded

Agree plan between parties

Pre-commencement checks
- loading/discharge line and tank valves correctly set
- correct hose(s) is fitted to appropriate coupling(s)
- loading/discharging hoses condition
- moorings effectively prevent strain on hose(s) from vessel movements;
- sufficient hose deployment to handle worst-case ranging of the vessel on the quay;
- communications routine between all parties is established;
- vessel's tank level monitoring system is proven.
- a watchman with suitable communication equipment is on duty at deck manifold;
- any relevant company safety check list or dangerous goods safety checklist are reviewed and signed-off
- for hyrocarbons, ensure:
  - Avery Hardoll connections used, not Weco type screw couplings
  - scuppers are plugged and Hot Work Permits withdrawn
- pollution prevention equipment is in place as per SOPEP
- drip trays, fire hoses, fire extinguishers, gas concentration test equipment and explosimeter are provided at vessel hose/manifold/reception site
- "No Smoking" conditions are imposed
- harbour/base operator's emergency pollution procedures displayed

Prepare and inspect bulk tanks

A deck officer or separate surveyor must ensure, and log, that tanks are ready for product and quantity to be shipped. Check:
1. Clean and empty
2. Air distribution slides and tank access seals in good condition
Shipper double-checks bulk tanks on board are prepared before commencing loading

Request information

Connect hoses and set lines
- type and quantity of product to be discharged and order of discharge
- estimated pumping rate for each
- length of warning/estimate of time required to stop
- whether stop will be initiated from vessel or installation/port operations
- established emergency stop procedure
- confirmation liquid bulk lines can be drained back to the vessel's tanks by gravity only

Start pumping at slow rate.

- vent lines connected and line + tank vented to atmospheric pressure

All lines & connections OK?

Cease pumping   No

Yes

- Check all system manifold connections for leaks.
- Check product entering correct tanks.
- For dry bulk transfers, use air purge first to clear lines, prove connections and ensure a good vent is obtained.

Fix problems

- During liquid loading manually check the level monitoring system, especially loading to near capacity of individual tanks.
- Pumping pressure of fuel and glycol may not exceed 7 bar. No more than requested quantity of fuel oil / glycol should be discharged

Full delivery rate

Reduce loading rate to top off tanks safely   Tank near full

Completed

Set lines to drain back to vessel's tank (except for potable water)

Use vacuum breaker if fitted to aid draining.
Where suitable use crane to raise hose to aid draining.
For dry bulk, use purge air to clear line.

When hose disconnected fit suitable cap / blank

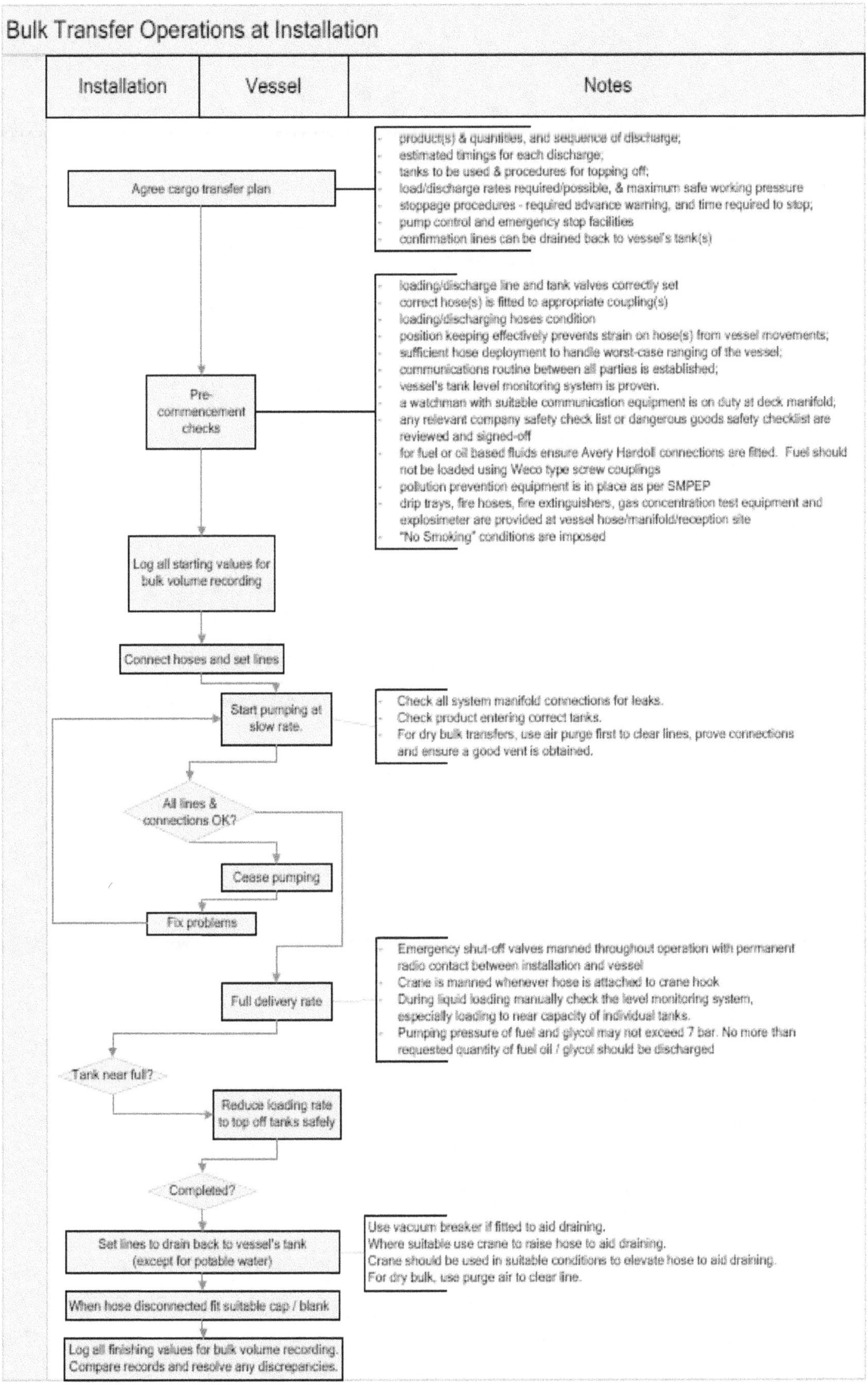

Bulk Transfer Operations at Installation

Installation	Vessel	Notes

Agree cargo transfer plan

- product(s) & quantities, and sequence of discharge;
- estimated timings for each discharge;
- tanks to be used & procedures for topping off;
- load/discharge rates required/possible, & maximum safe working pressure
- stoppage procedures - required advance warning, and time required to stop;
- pump control and emergency stop facilities
- confirmation lines can be drained back to vessel's tank(s)

Pre-commencement checks

- loading/discharge line and tank valves correctly set
- correct hose(s) is fitted to appropriate coupling(s)
- loading/discharging hoses condition
- position keeping effectively prevents strain on hose(s) from vessel movements;
- sufficient hose deployment to handle worst-case ranging of the vessel;
- communications routine between all parties is established;
- vessel's tank level monitoring system is proven.
- a watchman with suitable communication equipment is on duty at deck manifold;
- any relevant company safety check list or dangerous goods safety checklist are reviewed and signed-off
- for fuel or oil based fluids ensure Avery Hardoll connections are fitted. Fuel should not be loaded using Weco type screw couplings
- pollution prevention equipment is in place as per SMPEP
- drip trays, fire hoses, fire extinguishers, gas concentration test equipment and explosimeter are provided at vessel hose/manifold/reception site
- "No Smoking" conditions are imposed

Log all starting values for bulk volume recording

Connect hoses and set lines

Start pumping at slow rate.

- Check all system manifold connections for leaks.
- Check product entering correct tanks.
- For dry bulk transfers, use air purge first to clear lines, prove connections and ensure a good vent is obtained.

All lines & connections OK?

Cease pumping

Fix problems

Full delivery rate

- Emergency shut-off valves manned throughout operation with permanent radio contact between installation and vessel
- Crane is manned whenever hose is attached to crane hook
- During liquid loading manually check the level monitoring system, especially loading to near capacity of individual tanks.
- Pumping pressure of fuel and glycol may not exceed 7 bar. No more than requested quantity of fuel oil / glycol should be discharged

Tank near full?

Reduce loading rate to top off tanks safely

Completed?

Set lines to drain back to vessel's tank (except for potable water)

- Use vacuum breaker if fitted to aid draining.
- Where suitable use crane to raise hose to aid draining.
- Crane should be used in suitable conditions to elevate hose to aid draining.
- For dry bulk, use purge air to clear line.

When hose disconnected fit suitable cap / blank

Log all finishing values for bulk volume recording. Compare records and resolve any discrepancies.

Other Titles

Other related titles available from Magellan Maritime Press include:

- Introduction to Container Ship Operations
- Introduction to Maritime Safety
- Introduction to the Principles of Maritime Navigation
- Marine Engineering Systems
- Merchant Navy: An Introduction
- Hazard and Risk Analysis

Our titles are available for order from www.magellanmaritimepress.com and all good quality bookstores..